Human Cloning

Human Cloning

Playing God or Scientific Progress?

Lane P. Lester, Ph.D.
with
James C. Hefley, Ph.D.

Fleming H. Revell
A Division of Baker Book House
Grand Rapids, Michigan 49516

© 1998 by Lane P. Lester and James C. Hefley

Published by Fleming H. Revell
a division of Baker Book House Company
P.O. Box 6287, Grand Rapids, MI 49516-6287

Some chapters are taken largely from *Cloning: Miracle or Menace?* published in 1980 by Tyndale House.

Printed in the United States of America

Library of Congress Cataloging-in-Publication Data

Lester, Lane P.
    Human cloning : playing God or scientific progress / Lane P. Lester, with James C. Hefley.
        p. cm.
    "Some chapters are taken largely from Cloning, published in 1980 by Tyndale House."
    Includes bibliographical references (p.).
    ISBN 0-8007-5668-1 (pbk.)
    1. Human cloning—Moral and ethical aspects. I. Hefley, James C. II. Lester, Lane P. Cloning. III. Title.
QH442.2.L474   1998
241'.66—dc21   98-23847

For current information about all releases from Baker Book House, visit our web site:
                    http://www.bakerbooks.com

Contents

1

A Brave New World

People said some things would never happen.

A steam-driven boat chugging up the river without oarsmen.

A fuel-driven buggy zipping along at the incredible speed of thirty miles an hour. Even if some crazy did invent such a buggy, it would shake the intestines out of a rider.

An "airplane" made from cloth, wood, and wire, lifted into the air by a gasoline-powered engine.

A "gun" that shoots a missile ten thousand miles, hits a yard-wide target, and kills thousands of people.

A human walking on the moon.

A spaceship landing on Mars.

Molecular scientists building a human embryo outside the womb by fusing sperm and egg in a dish.

The cloning of a bull, a sheep, or a monkey.

All of the above have happened, most of them in our lifetime, shades of Aldous Huxley's futuristic *Brave New World* and David Rorvik's *In His Image*. Huxley (1894–1963) believed science was destroying human values. His satirical novel pictured a totalitarian society that tramples on human dignity and worships only science and machines. Huxley never suggested that his story was based on fact, unlike Rorvik, an award-winning science writer who formerly worked for *Time*. Rorvik claimed a male clone had already been secretly birthed. *In His Image* tells how an eccentric millionaire paid to have himself cloned by a master scientist. The respectable J. B. Lippincott Company, which published the book, said in a publisher's note: "The account that follows is an astonishing one. The author assures us it is true. We do not know."[1] (Rorvik's technical information appears quite accurate, but he used fictitious names for his principal characters and offered no external evidence of the actual cloning. On the basis of what he included in his book, I classify his story as a clever fake.)

That was then, but this is now: Scottish scientists recently took a big step forward toward cloning humans when they reproduced a female sheep in the exact genetic image of its "mother." No father needed, thank you. They named the newborn ewe Dolly after the country entertainer Dolly Parton.

Will cloning an actual human being be next? On a scale of one to ten, I'd say that scientists are at 9.9 in cloning a human—unless a workable government moratorium is put into effect against would-be cloners by the United States and United Nations. Even if this could be done, rogue governments such as North Korea could allow human cloning just for the money and scientific prestige it would bring. Iran and Iraq would ordinarily fit this category; however, they and other Muslim governments are unlikely to countenance cloning because of recent pronouncements by Islamic scholars at a world symposium. The experts there urged Muslim countries to ban human cloning but allowed work to go forward on the cloning of animals and plants. A spokesman for the Islamic Educational, Scientific, and Cultural Organization (IESCO) said at the closing session: "The symposium recommends the ban of . . . human cloning unless exceptional cases occur in [the] future."

The History of Cloning

Cloning is not new to biologists, although before Dolly the process was not on the front burner in genetic studies. Thirty or so years ago, it was mainly associated with horticulture and agriculture. The word *clone* refers to an organism or a group of organisms produced from a single organism or cell so that all have identical heredities. In general, the genetic manipulation that is called cloning has moved from lower forms of life to mammals.

The first successful freezing (at −79°C) of bull semen was announced in 1950. The procedure was then developed for the insemination of cows to produce better calves. Two years later, in 1952, the first actual animal cloning took place when frogs were developed from asexual tadpole cells. In 1962 other frogs were cloned using cells from older tadpoles.

In 1978 the gripping film *The Boys from Brazil* unfolded a chilling plot to clone little Hitlers. This same year, a child named "Baby Louise" was conceived in a laboratory dish through in vitro fertilization of a husband's sperm and his wife's egg. Two Englishmen, Dr. Patrick Steptoe and Dr. R. G. Edwards, attended the procedure from conception through birth. It was during this same year that David Rorvik's book *In His Image* created a mild sensation.

The year 1983 marked the first time an embryo was transferred from one human mother to another. Two years later the first transgenic livestock were announced. Scientist Ralph Brinster developed pigs that produced human growth hormone. The following year, Mary Beth Whitehead agreed to become a surrogate mother through artificial insemination. After Baby M's birth, Ms. Whitehead decided she wanted to keep the child, but the baby was given to its biological parents.

In 1993 television's *The X-Files* threw the spotlight on fictional psychotic clones, while *Jurassic Park* drew millions of movie fans with a film showing cloned dinosaurs. In the fact department, human embryos were cloned for the first time. The scientists took cells from seventeen defective human embryos that an infertility clinic had intended to destroy. They drew out the cells and sought to grow each one in a lab dish. They developed several thirty-two-cell embryos, each capable of being implanted in a woman's uterus and growing into a child. These scientists stopped the experiment at this stage and presumably destroyed the embryos.

In 1996 actor Michael Keaton cloned himself in the fictional movie *Multiplicity*. Truth then overpowered fiction when Scottish scientist Ian Wilmut and his team reported the sensational cloning of a sheep by using an adult cell. Most scientists agreed that the 1993 cloning of human embryos and the Scottish success had taken science a giant step forward toward human cloning.

The barrier to adult mammal cloning has been the way that, during development, cells become fixed in their destinies: skin, muscle, nerve, and so on. Ian Wilmut and his Scottish team set out to change the working of key cells so that they would act differently than when brought into existence. The Scottish researchers took udder cells from a six-year-old, pregnant sheep. They developed these cells in lab dishes and im-

10

mersed them in strength-building nutrients. Then, at a carefully determined time, they cut back the nutrient to one-twentieth of what cells normally need to grow. Five days later the cells were at rest, quite at precisely the time in their life cycle when their genes would be most vulnerable to reprogramming. The genes were now ready to act on signals from the female reproductive ovum cells—signals that were expected to tell them how to make a lamb embryo. The outcome proved far from predictable. Only eighteen pregnancies resulting from the forming of an embryo were engineered out of 277 adult cells used in the experiment. And from those eighteen embryos, only Dolly was born alive. Actually, this was an epochal success. No one had ever done it before. The Scots had found a way to switch on all the genes needed to birth a lamb from a single adult sheep cell.

Coinciding with the cloning of a sheep in Scotland came news of the cloning of two rhesus monkeys in Oregon. Scientists at the Oregon Regional Primate Research Center developed the two monkey embryos by taking a cell nucleus (the nest of chromosomes) from each of eight cells in a primitive embryo. They inserted the nuclei into egg cells whose own nuclei had been removed. These eggs were then implanted into surrogate mothers through in vitro fertilization. They will be raised by surrogate monkey mothers and could live up to ten years, researchers said.[2] The monkey cloning, however, may not presage the cloning of humans as might the Scottish experiment. The basic reason is that primitive embryos were duplicated in the monkey cloning, while the sheep scientists worked with adult animal cells.

The world media reported the Dolly story in great detail during the second week of March 1997. Cover stories were the norm—not since the end of World War II had there been such detailed coverage of a single event.

The three major United States news magazines also reported the Dolly epic with obvious excitement.[3] The original story was printed in the much lauded English scientific magazine *Nature*. Popular magazines and scientific publications in the United States picked up the cue there,[4] and prestigious business journals saw enormous prospects for moneymaking spin-offs in new genetic enterprises. *Business Magazine*'s cover story was titled "The Biotech Century," declaring, "Cloning animals is just the beginning."[5] Religious publications were no exception. Running a tight deadline, the liberal *Christian Century* ran a major theological article and a detailed news report on the sheep cloning. Quoting ethicist Paul Ramsey, theologian Allen Verhey wagged a verbal finger at would-be human cloners:

> Because the sexual person is "the body of his soul as well as the soul of his body," procreation (and intercourse) may not be reduced either to mere physiology or to simple consent to a technology. Because of our embodiment Ramsey refused to reduce baby-making (or lovemaking) to a technical accomplishment or to a matter of contract. . . .
>
> If we would cherish children as begotten, not made, as gifts, not products, then we will not be hospitable to cloning.[6]

Speaking with the editors of the conservative evangelical newsweekly *World*, R. Albert Mohler, president of Southern Baptist Theological Seminary, sounded severe storm warnings for the future:

> This [cloning] treads very near the sin of blasphemy, because those who would tamper with the genetic code are acting as creator rather than creature. To [tamper with the genetic code] is to risk not only intentional genetic mutation and its disastrous consequences for the human race, but to risk the coming of a very sinful and

evil dark age in which human beings will seek to define themselves, clone themselves, improve themselves, and finally perfect themselves.[7]

After Dolly's cloning in Scotland, Wilmut and his team received requests for human cloning by two families. "One inquiry," noted Wilmut, "was from a lady shortly to lose her father and the other was from a couple who lost their daughters in an automobile accident." Wilmut made it clear that both requests had been rejected. "I can't think of any embryologist I know who would be interested in cloning a human being," he told a news conference at the American Association for the Advancement of Science.[8]

The Approach of Human Cloning

If, or rather *when*, a research team does set out to clone a human, they will probably follow the path of the Scots. The cloners will take a cell from an adult human and "trick" it into multiplying, completely bypassing the normal process of conception. The embryo will then be implanted into a woman's borrowed uterus, where it will develop into a human fetus and be born the exact genetic twin of its single "parent."

Each one of all the hundred trillion or so cells in a human body (except red blood cells, which lose their nuclei during development) contains your total genetic blueprint, right down to the shape of your ears, the color of your eyes, and the unique loops and whorls of your fingerprints. Theoretically, you could be cloned innumerable times if there were enough wombs—human or artificial—to carry your offspring to term. Geniuses and other gifted people could be cloned after death, since cells can live for hours after vital organs have expired. And it would be no trick to freeze a body to preserve enough cells for cloning a select person; a wealthy person or a genius might leave some of his or her cells in a laboratory culture and bequeath a grant to the lab for his or her cloning. I expect, however, the first human clone will come from the cells of a living donor.

How Will Cloning Affect Us?

Dr. James Watson, awarded the Nobel prize in medicine for discovering the double helical structure of DNA, believes the initial response to human cloning will be "one of despair. The concept of the child-parent bond, not to mention everyone's values about the individual's uniqueness, could be changed beyond recognition."[9] Undoubtedly, human cloning will unleash a torrent of questions that strike at the root of basic values cherished in our society.

- What will happen to our concept of uniqueness and individuality?
- Will donors and their clones suffer identity crises as a result of bearing the same genes?
- Will cloning destroy the two-parent family?
- Will the Christian community condemn cloning as a mockery of God's will for humanity? Will liberal religious leaders, who now endorse same-sex marriages and even homosexual ordination, stop at cloning?
- Most troubling of all, will clones be made in God's image and likeness? Will they be/have eternal souls?
- Will God hold them responsible for their sins? Will the "parents" of clones bear responsibility too?
- Finally, on what basis can and will professing Christians oppose cloning?

We can guess how society at large will regard cloning from the avalanche of publicity resulting from the cloning of Dolly the sheep. The first cloning of a human being will certainly be a media event of unprecedented proportion. We'll most likely be inundated by television specials, movies, talk shows, books, feature articles, and global chats on the Internet. I expect that legislation (which is already beginning to take shape) will be introduced in many countries to forbid further cloning of humans. But will governmental authorities be able to control secret labs? Will they even be able to locate them?

Though many Christians are just awakening to the startling possibilities of cloning, in biological labs and graduate schools, vigorous debate over its possibility and practicability has been going on for decades. Among those opposing it are Dr. James Watson and U.S. Senator Christopher Bond. Watson worries that human cloning could destroy civilization. And Bond doesn't "think we should be playing God in trying to produce humans by cloning." Senator Bond, a member of the Senate subcommittee that oversees spending for the Department of Health and Human Services, wants a government ban on federal grants that could be used to police cloning research.[10]

On the let's-do-it team are the disciples of Joseph Fletcher, father of "situation ethics." Fletcher, also a pioneer of "biomedical ethics," endorsed cloning when "the greatest good for the greatest number is served."[11] Only the end, he believed, justified the means—nothing else.

Prelude to Cloning

Scientists were talking about the possibilities of human cloning for years before the media took on the subject in a big way. The first big story to grab public attention occurred on July 25, 1978, when the two English doctors, Steptoe and Edwards, announced the birth of the world's first "test-tube" baby. The headline-making report fascinated millions who wondered how a baby could be conceived in a test tube. Was this a prelude to cloning?

Briefly, this is what happened. Twenty-nine-year-old Lesley Brown had blocked oviducts, which prevented an otherwise fertile woman from conceiving through sexual intercourse. Like many other women unable to bear a child, Mrs. Brown was referred by her doctor to a specialist. "Don't worry," Dr. Patrick Steptoe told her, "you're just the right age for my purpose."

Medical researchers had been working on the test-tube procedure for several years. In 1974 Dr. Douglas Bevis told the British Medical Association that he had achieved the birth of three babies by IVF (*in vitro fertilization,* meaning "in glass"—outside the human body and in an artificial laboratory environment). When Bevis refused to identify the three babies and give specific details of the accomplishment, his story was disbelieved. However, Steptoe and his partner, Dr. Robert Edwards, spared no detail and produced the Brown baby in line with a lucrative publicity contract that the Browns had signed with the *London Daily Mail.*

In their epochal achievement, Steptoe and Edwards surgically removed an egg cell from Mrs. Brown and placed it in a nutrient solution along with a sample of her husband's sperm. One of the sperm cells entered the egg cell and fertilized it; so, in a sense, the mother-to-be conceived. The fertilized egg (zygote) divided and developed into an embryo, which the researchers transferred from the lab dish to Mrs. Brown's womb. A little less than nine months later, she gave birth to a healthy baby.

A second IVF baby was born October 3, 1978, in Calcutta, India. Today thousands of babies have been conceived by the Steptoe-Edwards method. The two Englishmen solved an intermediate problem in cloning: How do you nurture a developing embryo and implant it in a woman's womb? They did it!

The next big step toward human cloning came when scientists actually "persuaded" a one-parent human cell to grow into an embryo in the lab. The embryo could presumably be planted into a carrier's womb to grow and be born.

Apart from cloning, the process by which test-tube babies are born has made possible some startling scenarios in human procreation. The "biological" mother can provide her egg for fertilization with sperm donated by the "biological" father, whom the mother has never met. A "bearing" mother can receive the implant and carry the child to term. Then the "adoptive" mother and father can take the baby home.

Steptoe and Edwards started the reproductive wheels rolling. No sooner was Baby Brown's birth announced than a London woman asked to bear her sister's child by an embryo implant. The sister could not give birth in a normal way. Surrogate parenthood is now being performed in many clinics, and artificial insemination husband (AIH—using the husband's sperm) is done routinely by some doctors. Artificial insemination donor (AID—using another man's sperm) is usually performed secretly.

Sperm banks have long been used in scientific animal husbandry. A single select bull may "father" a thousand calves in one year. The animal's semen is kept frozen in sperm banks until needed for insemination. The history of bioscience indicates that daring new experiments are first done using animals, then people. The use of frozen sperm is a prime example of this general rule of order.

Through the same technology involved in animal insemination, scores of human babies are conceived each year. Chapter 7 will tell you more about AID. Beyond AID and IVF lie possibilities of "parent stores," where a hopeful couple or single man or woman, heterosexual or homosexual, can start a family by picking up a choice frozen embryo produced on a fertility "farm" that is ready for implantation. Based on the way moral values have been declining in western society, such stores could be in operation under government licensing within our lifetime. Your neighborhood parent store could be located in a "medical park" next to a gynecologist's office for convenience.

A "Brave New World"?

Does all this sound like something out of Aldous Huxley's nightmarish *Brave New World*? You can find this futuristic book in almost every library now, though there was a time when it was banned from many public libraries. Huxley intended the book as a satirical fantasy of impersonal science. I doubt that he ever dreamed, when it was published in 1932, that his fictional prediction would one day become reality. Huxley portrayed a world called Utopia, run by humanistic scientists and social planners and ruled by the Great Being, Ford. In Utopia traditional monogamous marriage, parenthood, and religious rites are practiced only by "Savages" permitted to live outside the new world state on reservations. Citizens of Utopia conform to the ways of science and socialism from conception to death.

Brave New World opens with a group of students touring the Central London Hatchery and Conditioning Centre. They see new life in bottles being fertilized, gestated, "decanted" instead of born, and behaviorally conditioned in assembly line fashion. A single female egg to be used for the lower castes has its normal growth arrested but then produces from eight to ninety-six "buds." Each bud grows into a perfectly formed human embryo, which is then placed in a great bottle containing the simulated environment of a mother's womb. While the unborn is gestating, future intelligence and caste are determined by the controlled flow of oxygen into the bottle. Those intended for lower castes are given less oxygen. Members of the lowest caste, called Epsilons, are born virtually mindless. They wear black and are programmed to do menial labor. Khaki-clad Deltas are made a bit smarter to handle more complicated jobs. Gammas, dressed in green, are pushed another notch up the social ladder. Elite Betas form the upper crust. Alphas are the leaders of Huxley's daring new society.

Standardized people. Instruments of social stability. Strictly controlled population. As new made-on-order humans emerge from the hatchery, the dead are cremated at the Slough Crematorium, the chemical compounds of their former bodies made into fertilizer—nothing wasted, always socially useful. During a citizen's lifetime, every physical need is satisfied. With procreation handled by hatcheries, all sex is recreational. Everyone has everyone else sexually. Rape is banished from the human vocabulary. Promiscuity is a sacred commandment. The dwellers in Utopia hold songfests in which they sing "Bottle of Mine" instead of "Mother of Mine." In a mockery of Christianity they take Communion in the name of the Great Ford. Calendar years are numbered A.F.—"After Ford."

In 1946 Huxley wrote a new introduction to *Brave New World* in which he said, "Today it seems quite possible that the horror may be upon us within a single century."[12] Huxley died in 1963. Were he living today, he might agree with Dr. Paul Ramsey, long-time professor of Moral Theology and Ethics at Princeton Theological Seminary:

> There shall come a time when there will be none like us to come after us. . . . The Central London Hatchery will become a possibility within the next 15 to 50 years. . . . Philosophers, theologians, moralists, churches and synagogues do not have the persuasive power to prevent the widespread social acceptance of morally objectionable technological achievements if they occur.[13]

In 1970 Ramsey was thinking of genetic engineering, which makes AID and IVF seem like high school freshman biology. Genetic engineers propose nothing less than redesigning the blueprint of inherited characteristics that predetermine the physical makeup of every person. Genetic engineering holds long dreamed-of possibilities for curing more than twenty-five hundred known genetic diseases at their source. Hemophilia, cystic fibrosis, muscular dystrophy, Huntington's chorea, and other dread afflictions could be wiped out. Many types of cancer and heart disease, along with other crippling ailments—known to be genetically predispositioned—are also included within target range during the next few years.

Some even think that if industry and the environment can be cleaned up to eliminate cancer-causing agents, a life span of nine hundred years may not be unrealistic. A few promise that, barring accidents, people born in the coming century will live forever. (As this book was being written, the oldest known woman in the world died in France. Jeanne Calment was 122. A nursing home in Pensacola, Florida, now claims the title for resident Augusta Watts. She turned 121 on August 15, 1997. Christian Mortenson, a resident of a nursing home in San Rafael, California, is not far behind at 115. "As long as I can keep interested in my surroundings," he says, "I'll go on.")[14]

Predictions of living unbelievably long lives may not be as fanciful as we think. Not so many years ago, scientists discovered that the key chemical compound of genes is a nucleic acid called DNA—short for deoxyribonucleic acid. The DNA in the cell nucleus does its work by dispatching a messenger molecule to places outside the nucleus where protein molecules are assembled. The messenger is another nucleic acid, called RNA—ribonucleic acid. RNA relays DNA's instructions to the protein factory, telling it what proteins to make. Put together in the order specified, the proteins determine the color of our eyes as well as every other human characteristic.

Just as physical scientists are delving into subatomic particles that form the foundations of matter, so genetic engineers are tinkering with the chemical alphabet with which the language of life is written. Already they've synthesized and recombined the basic genetic chemical, DNA, in simple life forms like bacteria. Experiments have been going on for several years to counteract "bad" DNA in human cells with "good" DNA. We are already moving toward altering undesirable hereditary characteristics. How far will this go? Some say science will eventually be able to design humans as they can only be idealized today.

Frankly, the scientific community is divided over whether heredity-changing research should continue. More than a few researchers lie awake nights worrying that a fast multiplying killer microbe, developed in their lab from recombined DNA, might escape and infect multitudes of people. Almost a quarter century ago, when much less was understood than now, many scientists became so alarmed that they actually called a voluntary halt to DNA experiments. The moratorium was not lifted until 1975, when safety guidelines were drawn up and accepted. Many scientists still don't think the safeguards are foolproof, fearing that science will build a monster. Worriers point to mistakes of the past, such as the frightening thalidomide disaster. This drug was originally sold outside the United States as a tranquilizer and taken primarily by women to help cope with nausea during the early weeks of pregnancy. Thalidomide apparently wreaked havoc with the development of the unborn, resulting in hundreds of children born with deformed limbs—some with shortened arms, for example, that resembled a seal's flippers.

Many Christian intellectuals also fear the directions in which biorevolutionaries could take us. Almost all of the leading research scientists in the genetic field are admitted atheists or agnostics. We need to heed a lesson from the Renaissance, when many intellectuals reacted against the failures of organized religion and led the younger generations into the so-called Enlightenment. Possessed by secular ideologies, France, in particular, was plunged into a bloody revolution that almost destroyed the nation. Like Voltaire and other God-mockers of the eighteenth-century Enlightenment, many of today's biological scientists look only within themselves for ethical guidance. They reject the idea of a sovereign, involved God who created us and to whom we are accountable. For them, humans are alone in the universe, a product of mindless evolution with no hope except in collective human wisdom. They hope for a disease-free super race. This is man-made evolution at the ultimate—the ideal human as visualized by humanistic social planners. This new human sees religion only as a vestige of the primitive past (an approach taken by most professors in secular universities and in some historic church-related schools), a humanistic utopia beyond *Brave New World.*

22

Is this where humanity is heading? Is the biorevolution of these last years of the millennium setting the stage for a Ford-like dictator to arise and proclaim himself God over all? We don't know the answers to these questions—yet. For the present, the key question is: How should we respond as Christians to the bioethical genetic revolution, with its enormous potential for good and evil?

2

In *Whose* Image?

In the previous chapter, I touched on how close we are not only to cloning a human being, but also to realizing the grotesque world Aldous Huxley imagined in *Brave New World*. In this chapter I want to explore and confront some of the ideas and positions presented in David Rorvik's provocative book *In His Image*. As stated before, I believe this work to be a clever, thoroughly researched fake. His philosophical conclusions, however, are neither fictional nor isolated—they represent the earnest (and alarming) contentions of many in the scientific community. Therefore, it will be worth our while to think through the implications of the belief system he champions and the issues he raises in his book.

Heredity: The Family Link

After the smashing success of Alex Haley's *Roots*, the best-selling book on his family line, a fourteen-year-old girl came home and announced to her parents that Haley had committed suicide.

"Why would he do that?" asked her gullible father.

"He found out he was adopted," she replied with mock seriousness.

Alex Haley actually lived many more years, collecting royalties from his book. Reference librarians are still receiving genealogy requests from ancestor hunters inspired by Haley's moving story. The great interest in genealogy triggered by *Roots* shows that nothing has yet come along to replace the sense of identity and belonging that a family ancestry provides. Cloning, as I'll explain a little later, will drastically alter family lines and units that people have known since creation.

Columnist Maggie Gallagher asks, "What exactly is the family status of a clone, anyway? Is the baby you have cloned from your DNA your child or merely your twin brother? Does he inherit from your estate or from your father's? Do you have to adopt your clone, and if so, will courts accept that it is in a child's best interest to be raised by a father who is also a genetic twin? Sigmund Freud, call your office!"[1]

Gallagher raises some intriguing questions—questions we can begin to answer by examining the nature of heredity.

Every physical and mental trait you and I possess has been affected by the gene pool of our family heredities. Certain traits are more obvious among the immediate generation, but occasionally some oddity shows up that we can't blame on any ancestor we know. That's usually an inheritance from a more distant predecessor or produced by something other than heredity. Your family tree defies the cataloging of every branch. In ten generations you can count about 1,056 grandparents. It wouldn't take many more generations to reach a million grandparents. Imagine: Your eye color, nose shape, skin texture, and countless other characteristics are determined by genes passed through thousands of grandparents, clear back to Adam and Eve. This potpourri of traits makes up your unique genotype (hereditary blueprint). Unless you are an identical twin, there is nobody else in the world quite like you.

Of Clones and Twins

Would the experiences of identical twins give us some insight into the relationship between a person and his or her clone? Could we, genetically speaking, put clones and twins into the same category?

In a sense, yes. Identical twins are a type of clone because they share the same hereditary traits. They are produced by the division of the cell mass that grows from a single fertilized egg. Identical twins are fairly rare, occurring once in approximately 344 births in the United States. And we might indeed learn a lot about what clones will be like by studying such twins, especially twins that were adopted as infants by different families.

Take twins James A. Spring and James E. Lewis, for example. Born in Piqua, Ohio, on August 19, 1939, they were adopted a few weeks later by families living only forty miles apart. Neither family was told that their adopted son had a twin brother, and the twins did not meet until almost forty years later. They discovered some amazing coincidences. Both had taken law enforcement training. Both pursued hobbies of blueprinting, drafting, and carpentry. Both had married first wives named Linda and second wives named Betty. Both named their first sons James Allen. Both were six feet tall and weighed 180 pounds. And, interestingly, both twins had the same brain wave and heartbeat patterns. The results of all the psychological tests they took looked as if one person had taken the same tests twice.

Referring to an account by Ian C. Wilson and John C. Reece, David Rorvik cites the case of identical twins who experienced the same illness and even died near the same time. Twin sisters apparently displayed the same symptoms of mental illness and were put in a room together at a mental hospital. When their condition worsened, attending physicians moved them to separate rooms to try to save one of them. That first night they were separated, the stronger sister was found curled up in a fetal position, dead. A few minutes later her twin was also found dead, lying in the same position. One of the psychiatrists described them as similar to "two people with one brain."[2] Since I am a geneticist and not a psychiatrist or a parapsychologist, I can't explain this. I can only suggest that this offers us some preview of relationships between people and their clones.

Identical twins have the same genetic "serial number" stamped in every cell of their bodies, as will members of a clone. This can be lifesaving when one needs an organ transplant. Identical twins Chris and Craig Bowman, whose parents were our fellow church members when we lived in Orlando, Florida, had such an experience. Years before, Chris began experiencing chronic backaches. His doctors removed a growth from his back, but he did not recover well from surgery. A more thorough investigation led to the diagnosis of cancer of his lymphoid tissue. Chris's only medical hope was a bone marrow transplant to arrest the growth of the cancer.

The big problem with transplants, however, is rejection. The human body has intricate mechanisms that

28

enable it to recognize foreign matter, whether disease bacteria or a donated heart. The body's defenses attack the foreigner as an intruder. Doctors counterattack with drugs to suppress the natural defenses, but this makes the patient more vulnerable to infectious diseases. The battle is touch and go and often results in rejection. Then the surgeons must look for a new donor.

Craig came to Chris's rescue and donated some of his bone marrow. Since it was the perfect genetic match, Chris's antibodies regarded the new marrow as their own and did not fight the transplant. The cancer was stopped, and Chris was completely healed.

Chris had only one identical twin to provide help. A member of a clone may have many such "twins." In time of need, any one of them could provide a transplant of identical tissue to another member of the clone group. Because of this possibility, people may have themselves cloned just to insure having a supply of young donors available when they need new body tissue or a new organ. The ethical ramifications of such a scenario are startling.

The major difference between identical twins and members of a clone is their relative ages. Complicating matters is the fact that, just as twins share the same parents, so the cell donor and the clonal "child" have the same parents. The cell donor is no more the biological parent of the clone than is one identical twin the parent of the other. It would be more appropriate for a cell donor to relate emotionally to its clone as a sibling than as a parent.

Heredity and Environment

Nonidentical (fraternal) twins spring from the fertilization of two eggs by two sperm. Dr. Steve Jones, professor of genetics at the Galton Laboratory of University College in London, highlights the differences between fraternal twins through the biblical example of Jacob and Esau: "'Esau was a cunning hunter, a man of the fields, and Jacob was a plain man, dwelling in tents.' They looked quite different." Esau was a hairy man and even had "different ways of speaking: 'The voice is Jacob's voice, but the hands are the hands of Esau.'"[3]

Jones notes that identical twins are influenced by their environment, just as human clones will be. "The very fact of being identical twins," Jones reasons, ". . . sharing very similar names and dressed in identical clothes—may predispose to mental disease. Twins often have a poor environment before birth as they share a placenta. . . . Their similarity may be due more to a shared environment than as first appears." Such reasoning, Jones says, is backed by studies. "Identical twins are twice as likely to suffer from coronary heart disease than are those of a pair of fraternal twins, and five times as likely to have diabetes. Even tuberculosis is more commonly shared between identical than fraternal twins, suggesting that there may be an inherited basis for susceptibility."[4]

Obviously, studies on the upbringing of clonal humans have not been done—yet. If, and more likely when, humans are cloned, scientists will compete to study their characteristics and actions. Questions of nature versus nurture are certain to be among the many challenges considered. Wherever we position ourselves on future possibilities, it is absolutely essential that we understand the basis of human genetics as it relates to the replication of life.

Genes: The Heredity Link

We once thought that the cell, the basic unit of life, was a simple bag of protoplasm. Then we learned that each cell in any life form is a teeming micro-universe of compartments, structures, and chemical agents—and each human being has billions of cells. We call the hereditary, trainlike specks that swim in the cell nucleus *chromosomes,* meaning "colored bodies," from the way they absorb laboratory dye. The "passengers" they carry are genes, and genes are made of the vital chemical DNA.

Normal Sexual Reproduction

Every normal living thing has a specified number of chromosomes in every body cell. A corn cell has twenty; a mouse, forty; a gibbon, forty-four; and a human cell, forty-six. The forty-six chromosomes in a normal human cell come in two matching sets of twenty-three each, with each parent providing one set. The chromosomes are numbered one to twenty-three. To have more than two of a number is abnormal and almost always results in tragedy. In human fertilization, the father's sperm cell with twenty-three chromosomes unites with the mother's egg cell containing the same number. This produces a single cell with forty-six chromosomes that starts dividing. In a few days there is a discernible embryo. The hereditary blueprint, a product of the parents' heredities, was there from the instant of fertilization.

Unfortunately every new blueprint is flawed by genetic "mistakes" accumulated and passed down the family lines. Some mistakes produce one or more of the twenty-five hundred known genetic diseases, such as hemophilia, in which a critical blood-clotting substance is absent, leaving the victim in danger of bleeding to death from a slight injury or bruise. Hemophilia is hereditary, and it almost always affects males. However, mothers who show no sign of the malady can transmit the disease to their sons. In one family, a daughter did not have the disease, but one of her brothers did. She

31

chose not to marry because of the real possibility of passing on the disease to her future male children.

Genetic mistakes, which cause some of humankind's most dread diseases, are called *mutations*, alterations of the genetic blueprint. This may involve the abnormal duplication of a chromosome, as in Down's syndrome, in which there are three chromosome twenty-ones instead of the normal two. A broken portion of a chromosome may be lost (a deletion), or it may connect with an unrelated chromosome (called a translocation). In some cases, the chromosomes may all be normal, but a single gene on a chromosome may be inaccurately copied. The result of all these possibilities is "typographical errors" in copying the genes for the next generation.

Jesus' Unique Birth

Every human has developed from the genetic blueprint bequeathed by the combined heredities of his or her parents—except Jesus Christ, the virgin-born Son of Mary whose Father was God. Scripture says this plainly. The birth of Jesus was predicted by many Old Testament prophets. Isaiah declared: "Behold, a virgin shall conceive, and bear a son, and shall call his name Immanuel" (Isa. 7:14 KJV). The virgin birth was accepted as fact by the New Testament writers (Matt. 1:18, 23; Luke 1:31–35). In denying the virgin birth of our Lord, skeptics are calling the prophets and the apostles liars.

Because cloning is a sort of virgin birth, we need to consider the unique nativity that occurred in Bethlehem. Unbelievers, of course, say the biblical writers papered over the real story. One theory is that Mary was impregnated by a soldier from the Roman occupation army. Another says that Joseph and Mary succumbed to passion while they were engaged. Jesus' enemies, however, would surely have used either of these to refute Jesus' claim to deity.

Still another foolish speculation is that the virgin birth was a freak biological happening, similar to the parthenogenesis (Greek for virgin birth) that occurs in much of the plant and animal world. By the laws of genetics, parthenogenesis is impossible for human beings. Animal virgin births involve doubling the chromosomes in a female egg to give the egg a full number of chromosomes and the ability to develop into an adult. Through generation after generation, the human gene pool has accumulated a vast amount of genetic "garbage." To double the chromosomes in a human egg would result in a double dose of harmful mutations and would almost certainly be lethal. Even if the baby were to survive, he or she would be a genetic cripple, severely retarded and grossly deformed. Doubling through normal fertilization by a sperm usually does not result in mutations "stacked" one on another. The differences in the two heredities prevent this.

But even if it could happen, Jesus was not born this way, because he would have been a female. Follow closely, and I think you'll understand. One of the chromosomes in the human egg is always an X in genetic language. The male sperm cell may have either an X or a Y chromosome. If the sperm that fertilizes the egg carries an X, the resulting fertilized egg (called a *zygote*) will have two Xs and be female. If the sperm carries a Y, the zygote will be male. If Mary's chromosomes had doubled to produce Jesus, "he" would have had two X chromosomes and would have been a "she."

Then how could Jesus have been born without a human father? God could easily have created a special Y sperm, with twenty-three perfect chromosomes, to fertilize one of Mary's eggs. Since I accept God as being the Creator of the universe and of each "kind" of life, I have no difficulty believing that he used this method or some other means beyond the usual design for reproduction to accomplish this biological miracle.

God designed the true virgin birth; but in an effort to play God, scientists are creating their own "virgin births"—through cloning.

Cloning

Science has long known how to make simple life forms reproduce asexually. My first experience in this was as an undergraduate at the University of Florida, where, in a microbiology lab, we were easily able to produce clones of bacteria. In recent years cloning has been accomplished with higher life forms that do not normally reproduce asexually. Frogs and mice have been cloned, and more recently monkeys, a bull calf, and Dolly the sheep.

The first step in cloning frogs was to obtain a female ripe with egg cells. The eggs were removed with a skillful squeeze and then exposed to a strong light. This caused the eggs to lose their nuclei. (Remember, the nucleus of a cell, including an egg cell, contains the chromosomes—the carriers of heredity. A frog egg without a nucleus is therefore a cell with virtually no heredity, with no blueprint or instructions on how to develop into a frog.) The cloning scientist next removed a tadpole from his aquarium and cut out the intestine. He then extracted a nucleus from an intestinal cell, which he inserted into a nucleus-less egg. He repeated this process with other nuclei for the number of frogs he wanted to clone. The eggs with their new nuclei developed into tadpoles—all identical twins to the tadpole who gave his life for science. The tadpoles grew into adult frogs, each an exact duplicate of the others.

The procedure used to produce mice was different. Each mouse cell has forty chromosomes—with the exception of reproductive cells (mouse eggs and sperm have only twenty chromosomes). At fertilization, a twenty-chromosome sperm fuses with its egg counterpart to get the full complement of forty, just the way it happens in human reproduction when the twenty-three chromosomes from each parent come together. In the case of the mouse, rather than destroying the egg cell's nucleus and replacing it with another cell's nucleus, the scientist treated the mouse egg in such a way as to double the number of chromosomes in the egg nucleus itself. The treated egg with forty chromosomes was then placed back into the female uterus, where it developed into a baby mouse. Although this was certainly a departure from the usual way mice are produced, the new baby was not a true clone of the mother. An egg (or sperm) cell, you see, has a random sample of one gene from each of the many pairs of genes in the parents' cells. By being produced from duplicated egg chromosomes, similar to parthenogenesis, both members of the gene pairs in the baby were identical. Instead of the baby mouse possessing all of the mother's genes, as would be the case of a clone, it has only half the mother's genes in a double dose.

The true cloning of other creatures involved different techniques. The cloned calf was produced in the following manner, according to Michael D. Bishop, a vice president of research for the Infigen Company, a subsidiary for the ABS Global Company of Wisconsin. First, scientists took the genetic material from a cell that came from a thirty-day-old fetus and inserted it into an egg cell from which the genetic material had been removed. This produced a single cell that was an exact copy, or clone, of the cells in the thirty-day-old calf fetus.

The single-cell clone was provided with nutrients that would stimulate its growth and division into many cells. One of these cells was introduced into another egg from which the nucleus had been removed. This egg, with its new nucleus, then developed into an immature embryo for seven days and was then implanted into a recipient and gestated for 280 days until it was born. Michael Bishop says the calf, named Gene, came from one of eighteen embryos implanted into recipient cows, but only Gene was produced.

Near the announcement of Gene's cloning, a group of Japanese scientists said they had developed cloning technology that could produce up to two hundred identical cattle from a single fertilized egg. Their technique grows cultures of a fertilized egg, from which cell nuclei are taken and transferred to unfertilized eggs. An official of their company said cattle cloning technologies used previously have been capable of producing only three to five head of cattle from a single fertilized egg.[5]

Those who cloned monkeys in Oregon created two monkey embryos by taking a set of chromosomes from each of the eight cells in a monkey embryo. They inserted these into egg cells from which the DNA had been removed. The embryos were then implanted into surrogate mothers through in vitro fertilization. The monkey cloning had less dramatic impact than the cloning of the sheep, because primitive embryos, rather than adult embryos, were duplicated. However, this was the first time cloning was used to reproduce animals so closely akin to humans.[6]

Dolly, produced by Ian Wilmut and his team, came from an udder cell of an adult six-year-old sheep. Michael Bishop admits that cloning from adult cells is more difficult because the cells have already "specialized into specific cell types, such as skin, muscle, and bone."[7]

The method used in frog cloning may aid scientists intent on human cloning. Researchers persuaded women patients having abdominal surgery to donate eggs. They extracted the nuclei from the eggs, leaving the rest of each egg intact but robbed of virtually all its heredity. But instead of using intestinal cells as donor cells the scientists collected special cells from male patient volunteers having surgery on their sex organs. These cells are "parent" cells for sperm cells. In the male they are extracted from the testes. Each bears a full order of forty-six chromosomes. It is the sperm cells and egg cells that contain only twenty-three.

Dr. Landrum B. Shettles, one of the foremost biological scientists of modern times, has reportedly made substantial progress with sex cells. Three times he reports successfully removing the nucleus from a female egg cell and inserting the nucleus from a parent cell. Each time, the egg, with its implanted nucleus of forty-six chromosomes, grew and divided until it reached the multicelled stage, when, according to Dr. Shettles, it was ready to be implanted into a woman's uterus using the process perfected for test-tube babies. Dr. Shettles claims he halted the experiment at this point. I presume this means he killed and preserved the human cell masses for future study. He said, "There was every indication that each specimen was developing normally and could readily have been transferred into a uterus . . . where it would have developed into an identical twin of the man who donated one of his sex cells."[8] If Shettles accomplished what he claimed, his work may help many would-be human cloners.

A major obstacle in human cloning involves the genetic destiny of body cells. Some cells are destined to build fingernails, some to construct nasal passages, and so on. Yet each cell continues to hold forty-six chromosomes with blueprints for the whole body. The sex cells that Shettles claims to have used in his experiments, however, are not limited in their destiny, because their offspring—sperm cells and egg cells—must be capable of producing all kinds of cells. The use of body cells that have already been "switched on" for special jobs may not be successful in human cloning. However, the Scottish scientists who cloned Dolly may have found a way to break through this barrier.

Cloning Is Not Just Possible—It's Probable

What do others in the scientific community think of cloning's possibilities in light of recent animal cloning? Dr. Kurt Hirschhorn, chief of the division of medical genetics at Mount Sinai School of Medicine in New York City, is a former president of the American Society of Human Geneticists. He thinks human cloning is not only possible but probable. Dr. Bentley Glass, distinguished professor of biology at the State University of New York, projects "C-Day" by the end of this century. Dr. James F. Bonner of the faculty of the California Institute of Technology said as far back as 1968 that within fifteen years science would know how "to order up carbon copies of people." Science writer Isaac Asimov, now deceased, said before the recent accomplishments, "Biologists are sure that some day they will be able to clone mammals, and even human beings."[9] If Asimov were to come back from the dead today, he'd probably look at the optimism generated by the cloning of an adult sheep and say, "See, didn't I tell you?"

It's interesting that David Rorvik and Dr. Landrum Shettles, the scientist who claimed to have developed a clonal embryo, were good friends. A quarter-century ago they coauthored a book on how to choose your baby's sex before birth, and Rorvik used Shettles as a source for many of his articles. In *In His Image,* Rorvik claimed that the first human was cloned in December 1976. I don't believe him. He shrouded his experiments in mystery by giving made-up names to his characters and by being less than totally specific about the methods used. The friendship of Shettles and Rorvik, and the fact that Shettles claimed to have partly succeeded, makes me think that Shettles may have been the mysterious Dr. Darwin of Rorvik's book—although Rorvik stated at the time that Shettles was not the scientist. The name Darwin is a key to the evolutionary humanism that is behind Rorvik's 1978 book and the research into cloning as well. Rorvik decided moral questions by consequentialist ethics, another way of saying that the end justifies the means. Using this reasoning, Rorvik justified Darwin's deceiving the women used in the experiments, not telling some that eggs had been taken from them and not telling others about being impregnated with the cloned cell.

Even if Rorvik's story and other mind-boggling stories of human cloning are fakes, I still predict that human cloning is coming soon. Many different lines of research are converging to give us the capability of making genetic copies of ourselves. Just as Hitler's doctors, using consequentialist ethics, were able to justify heinous experiments for useful information, Rorvik and others have shown that human cloning will be done if the right people want it to happen.

Should It Be Done Simply Because It Can Be Done?

Human cloning will not be attained without much trial and error and the destruction of numerous egg cells and human embryos. Some of the embryos may develop into cloned "monsters" before they die or are killed in the laboratory. Steptoe and Edwards admit many failures before producing the first test-tube baby. Others who have worked on IVF embryo transplants have admitted dumping the remains of their experiments into the laboratory garbage for burning in an incinerator. Rorvik confessed in *In His Image* that his "worst doubts about cloning concerned the issue of 'bench embryos,' ... experimentally conceived in the laboratory and then sacrificed in the process of carrying out some research procedure."[10] Apparently Rorvik overcame his qualms by use of consequentialist ethics.

Many scientists during and after Rorvik's time have taken strong stands against human cloning. "The potential for human cloning," said Jeremy Rifkin, coauthor of the book *Who Should Play God?*, "no matter how far in the future, challenges our entire value system. We must talk about the implications now, before any crisis occurs."[11] Embryologist Robert T. Francoeur, a world-renowned scientist and author of *Utopian Motherhood,* has announced his opposition. "Xeroxing of people?" he asked. "It shouldn't be done in the labs, even once, with humans."[12]

Ian Wilmut and his Scottish colleagues told a Select Committee of the British House of Commons that the nuclear transfer technique they used to produce the sheep could in theory be applied to the cloning of humans. "Whether any scientist would try and whether it would work is another matter," says Dr. Donald Bruce, a representative of the Church of Scotland's Society, Religion, and Technology Project. According to Bruce, the Wilmut team at the Roslin Institute, where the sheep cloning was accomplished, "have made it quite clear that they think that to clone humans would be unethical."[13] Bruce adds that some

> have speculated whether it would be possible, on the basis of these discoveries, to grow not an entire being,

but living organs from cells. This might have certain potential to treat diseased organs or malfunctioning body
processes. There ... might be less of a problem ethically.
In that sense it might be no more of a problem than the
concept of an artificial heart, toward which some
progress has already been made.

Bruce thinks that the most critical problem might be

"gradualism." By a progression of small steps you could
eventually provide all the conditions needed to clone the
entire human being, even though it has never been the
intention of the research. This raises a much deeper
question about how the direction of research is [to be]
determined and controlled.[14]

The Scottish clergyman needs to be heard. Cloning
excepted, we have witnessed a steady erosion of concern for the core of the human species since experiments
began on the first stages of life. Who speaks today for
the embryonic lives that were destroyed to perfect the
techniques used to bring about the birth of test-tube babies? If these embryos can be extinguished for the sake
of science, what is wrong with extinguishing life at any
stage?

The recent trio of animal clonings (calf, sheep, and
two monkeys) provoked an unprecedented reaction and
a strong demand that scientists cease any research un-
derway on the cloning of humans. President Clinton announced a ban on using federal funds for human cloning
and appealed for a voluntary moratorium on any private research on the act. "Any discovery that touches
upon human creation," Clinton said, "is not simply a
matter of scientific inquiry. It is a matter of morality and
spirituality as well."[15] The Vatican called for a ban on
cloning humans, with Pope John Paul II denouncing
"dangerous experiments that harm human dignity."[16]

Conservative Protestants, committed Catholics, and Orthodox Jews all have taken strong stands against human cloning. Many others have also.

Only a minority of notables have said that human cloning could accomplish much good. "I welcome it," stated Senator Thomas Harkin of Iowa. "I don't think there are any appropriate limits to human knowledge. What utter, utter nonsense to think that somehow we can hold up a hand and say, 'Stop.'"[17]

The issue of cloning, however, has problematic implications for those of us who believe all human life is sacred at all stages of development. We will consider these in the next chapter.

3

Will Clones Have Souls?

A handsome young man we'll call Bill asks to join your church. Bill never had a mother. Because there was no mixing of genes from two heredities at his conception, Bill is his "father's" twin. He is truly a "chip off the old block." He not only looks like his father; in a hereditary sense, he is one with his father with no substantial genetic variation.[1] Bill is a clonal man. With Bill's request for membership, your church faces the question: Do clonal beings bear the spiritual "image of God"? Do they have the capacity to know and worship God? Will they, like ordinary human beings, be sinners by nature and by choice? Or will they be amoral, neutral, and not responsible for their own acts?

These questions do not concern those who see humanity as only a happening of cosmic chance that evolved from a bubble of amino acids into mechanistic

43

life. But they do challenge those who perceive human beings as more than chemistry and biology. Richard Land, president of the Southern Baptist Ethics and Liberty Commission, states well what I believe about human life:

> At the heart of the Judeo-Christian religious moral tradition is the absolute belief and conviction that human beings are made uniquely in the image and likeness of God. This belief in the stamp of the divine image on human beings is what sets them apart from animals. Animal life deserves respect; human life demands reverence.[2]

Souls and Sin

Humanity is soul and spirit, we believe, made in the image of a loving, all-powerful Deity. Though marred by a sin nature that expresses itself in rebellion against God, self-seeking, and self-exaltation, we are sought by God through the atoning, redemptive sacrifice of Jesus Christ. So we must ask now: Will Bill and other clonal offspring be subhuman, lower than ordinary Homo sapiens but higher than apes? Will God love clonal Homo sapiens as much as he loves those who come into the world the natural way?

We must think about where clones will stand in the order of God's creation—and we must think about it now. We cannot afford to be caught unprepared when cloning does happen. Rather than leaving the field to those who do not value God, we should begin now to shape the ideas and the ethics of the future. The spiritual dignity and value of humanity in Western society could be at stake. With the Bible as our basic sourcebook, let's begin in the Genesis record, which is neither fable nor "true myth" but authoritative and true.

The Genesis Record

The focus in Genesis 1 is on the entire creation process. God is Planner, Architect, and Contractor. Earth and seas, herbs and trees, fish and fowl, animals, and finally man, the crowning touch, are created at his command. He created all life forms by "kinds." From these kinds came the species that we know today. The cat "kind," for example, may include everything from tabbies to tigers, and the horse kind may range from Shetland ponies to zebras. The original members of each kind must have carried a great deal of genetic variety. As each kind obeyed God's command to "be fruitful and multiply," the processes of reproduction produced within each kind a number of species.

Scientists who accept the creation account of Genesis hang creatures by kinds on separate ancestral "trees." Evolutionists, on the other hand, drape all living forms on one big tree and claim that all evolved from the accidental explosion of life from dead matter at the tap root. (Incidentally, I was a Christian evolutionist—believing God was behind the evolving—until after receiving my doctorate in genetics. On being challenged by Dr. Duane T. Gish of the Institute for Creation Research, I took a year to sort out the evidence. I concluded there is more scientific reason to believe in special creation than in evolution.)

The second chapter of Genesis focuses on man and the environment that God prepared for him. The key statement is Genesis 2:7: "Then the LORD God formed man of dust from the ground, and breathed into his nostrils the breath of life; and man became a living being." This verse can be interpreted two ways. One is that God breathed animate physical life into Adam. The Hebrew word for "being" (translated "soul" in the King James Version) is *nephesh*. It is used in some other Old Testament passages for the life principle essential to the ex-

45

istence of animals and human beings. The second interpretation is that God breathed into man an eternal spirit, which animals do not possess. This belief comes from the fact that God breathed life only into Adam, but he commanded, "Let the earth bring forth" for the animals. This second view is more convincing to me. God is Spirit, and the life he breathed into Adam was spiritual life.

Scripture speaks of the human spirit hundreds of times. The word has different meanings in respect to contexts, but in general the connotation is that our spirit is distinctive and unique among God's creatures. *Spirit* is the moral and spiritual essence that is in God's image and is not limited by the physical form, which returns to the dust of the earth at death. Spirit is the channel through which God communes with humans. So Jesus told the Samaritan woman, "God is spirit, and those who worship Him must worship in spirit and truth" (John 4:24). No animal has the spiritual capacity to know and worship God. Will a clonal offspring, formed in the genetic image of a single human parent, be a spiritual being? The creation of woman may give us some hint.

> So the LORD God caused a deep sleep to fall upon the man, and he slept; then He took one of his ribs, and closed up the flesh at that place. And the LORD God fashioned into a woman the rib which He had taken from the man, and brought her to the man.
>
> Genesis 2:21–22

Was this the first cloning, since Eve was made from a single parent? By the laws of genetics as we know them today, she could not have received her total genotype from Adam, or else she would have been a male. Clonal reproduction always produces the same sex.

We've said that a male is XY and a female XX in genetic parlance. Using Adam's rib as the raw material,

God may have eliminated the Y chromosomes and duplicated the X chromosomes in each cell. This would be the minimum genetic reconstruction required. More likely, God completely reorganized the genetic molecules into a brand-new genotype for Eve. In any case, Eve was created before the reproductive cycle by sexual union began, with its mixing of parental genotypes to produce new genetic identities. This leads to a thornier question: Did Eve receive her *imago dei*, her spiritual identity, through Adam, or was she given her spiritual identity during God's "surgical" procedure? Since God didn't tell us in Genesis, I can't answer that. If the former is true, it could indicate that each of us is "stamped" with God's image through or from our parentage when we come into physical existence. A clonal person, then, would bear the image of God as much as anyone else.

Will a clonal person be responsible to God for his or her moral failings? Is he or she included in God's plan of redemption? Again, let's look at the first couple. Adam and Eve were created perfect. Not one of the thousands of genes in the forty-six chromosomes of each of their billions of cells was blemished by a harmful mutation. Their environment was also untainted by the pollution that plagues us today. Their bodies would have suffered a lot of wear and tear, but in their primeval perfection the life processes may have followed a renewal cycle that would have enabled them to live forever. However, they listened to the tempter and disobeyed God by eating the forbidden fruit. God had said, "You shall not eat from it or touch it, lest you die" (Gen. 3:3). This meant spiritual death, and physical death as well, if God had originally intended that their bodies last forever.

The immediate results of the first sin are presented in the Genesis account of the fall. Both Adam and Eve felt guilty and tried to hide from God. They were cast out of Eden. Satan, who took the form of a serpent, was told there would be "enmity" between his seed and Eve's seed: "He [the Seed of the woman] shall bruise you on the head," God informed the serpent, "and you shall bruise him on the heel" (Gen. 3:15). Many theologians call this the first Messianic promise, a foreshadowing of the conflict between Jesus and Satan, culminating in Jesus' ultimate victory over the devil. Eve will bear children in "sorrow" (the pain of childbirth), the Genesis account predicts. The man will "rule" over her (women have been exploited by men ever since). The ground under them is "cursed." Adam must sweat to make a living among the thorns and thistles until he returns to the ground in death. This indicates that if Adam and Eve had not sinned, their physical bodies might have endured forever.

Sin spread through Adam and Eve's posterity. "Therefore, just as through one man sin entered into the world, and death through sin, and so death spread to all men, because all sinned" (Rom. 5:12). Scripture characterizes sin as the universal malady of the human race, involving transgression, going astray, missing the mark set by God, and willful rebellion (see Isa. 53:6; Rom. 3:10, 23). The only remedy is the atoning, substitutionary death of Christ. God "made Him who knew no sin to be sin on our behalf, that we might become the righteousness of God in Him" (2 Cor. 5:21).

There is no reason to believe that clonal people will not be sinners, just as we are. I don't think there are any genes for sin, yet it is a fact that sin flows from one generation to another. Learning to sin from association with others is not sufficient explanation. Kids grow up doing what comes naturally—sin. What is more natural than sin? Clones will sin like everyone else. They will be in need of redemption. This should be reason enough to include them in God's circle of redemptive love and accept those who believe into our congregations.

When Does a Person Get a Soul?

Back to the matter of spiritual identity. When will a clonal life become a person with unmeasured value in God's sight? How does spiritual identity arrive? What or who is the conveyor? Is it borne along through the reproductive channels of the generations? Is it related to heredity? Are two parents necessary? These are sufficient questions to keep Christian theologians and philosophers debating for a thousand years. I don't have any final answers. I can only say that every individual has a self-identity; one is always a complete being, composed of body, spirit, and soul. You are never a half person. Genetic likeness will not merge two into one.

Exactly when during the conception-implantation-gestation-birth process does one become a soul? This is relevant to cloning, test-tube babies, and other new methods of reproduction and childbirth. It is critical in the abortion controversy. The Catholic Church officially holds that the soul or spirit is infused at the moment of conception. Original sin, Vatican authorities state, is also present from this instant. Thus Catholic dogma outlaws IUD "birth control" devices and morning-after pills that destroy a fertilized egg. As you might expect, Protestant theologians and doctors disagree among themselves. The most conservative agree, with theologian Francis Schaeffer and pediatric surgeon C. Everett Koop, that personhood is attained at conception. Dr. Thomas Monroe, an obstetrician-gynecologist and Presbyterian lay leader in Chattanooga, and others take a more modified stance. They believe the soul is given when the fertilized egg has "slid" down the mother's oviduct, developed into an embryo (blastocyst), and implanted itself in the wall of the uterus. The position of pro-abortion liberals is that the new life is not a full person until born with the capacity to sustain itself. I wonder what the liberals think about preemies who cannot survive independently but can be sustained by modern medical technology.

Personally, I think the implantation argument is an escape hatch for allowing early abortion. Biological life truly begins at conception, since all the genetic blueprints for the baby's future development are in the fertilized egg. I can't see how implantation makes a substantial difference, except that the new life has a much greater chance of succeeding. Is it really any different from the life in the fertilized egg? I think not.

However, we run into some hard questions on cloning if we say that a person exists spiritually as well as physically at the moment of conception. Identical twins, which are a type of clone, develop not at conception but a number of cell divisions later, when the mass of cells breaks apart to form two identical and separate cell systems. If the spirit was given at conception, then the twins might have to share a common spirit, or one might do without a spirit altogether. A clonal offspring, you remember, is also an identical twin to its "parent."

Furthermore, there is at least one documented case in which two fertilized eggs developed and fused to produce one person. The last I heard, this woman is still living in Vienna, Austria. Scientists at the University of Vienna, who discovered the unusual birth, call her a genetic mosaic. Her parents produced two fertilized eggs, which happens with fraternal twins, but early in development the embryos merged to form one fetus. Cells from one of the fertilized eggs made the woman's blood and skin cells, while cells from the other eventually produced her eggs.[3] I expect that this differentiation may have happened elsewhere, but the phenomenon went undetected. When two fertilized eggs become one, is there a doubling of the spiritual dimension in one person? We're edging toward theological absurdities. I'm satisfied to leave these unexplainables to the wisdom of

God. If we knew everything, we would be God! We just don't know precisely how or when a human becomes a spiritual being.

God may not work by our rules at all. It's that way in physics. Dr. John H. Martin, a researcher at Argonne National Laboratory near Chicago, observed that an electron in the nucleus of the atom can be made to go through two different places at once. "This is impossible in our familiar world," he noted. "I know that my boy playing baseball cannot bat a fly into both right and left fields at the same time." Yet an electron can go two places at once! What does a physicist do when he meets the "unreasonable"? Dr. Martin said, "We always assume . . . that it is our understanding that is at fault."[4] Biologists should be just as reverent.

Though we can't pinpoint the moment an embryonic human receives spiritual identity, this doesn't mean that a human is not a spiritual being. There are many evidences that humans are vastly different from apes, orangutans, dolphins, and other "smart" animals. If we can't be so definitive about humans conceived and born according to the ordinary laws of nature, I doubt that we'll do any better with cloned humans. Given our deficiency of wisdom, I think we should be prepared to award full personhood to any cloned person who comes to be.

We won't be able to tell clonal people from people born the ordinary way. From what we know about animal cloning, a cloned human will have all the human characteristics of speech, thought, senses, and physical appearance. The clone will be as different from an ape, for example, as you and I are.

Whole Human Beings

If I were on the church membership committee, I would vote for Bill to be accepted into the local Christian fellowship if he professed faith in Christ. I would accept him on the same basis as I would anyone else. The fact that he is both the son and twin of his "father" should have nothing to do with his relationship to me and to our Christian brothers and sisters. Bill is as human as my own son.

All of the Bible-believing professors and theologians whom I have talked to or read are troubled over the prospect of human cloning. David Gushee, professor of ethics at Union (Baptist) University in Jackson, Tennessee, says: "It is clear to me from a Christian perspective that under no circumstances of any kind would it ever be morally permissible to clone a human being." Gushee, as I do, finds in human history a recurring pattern among people—our ability to make use of every scientific and technological innovation to commit bodily harm, as well as to help one another. He cites the history of warfare as an example.[5] Gushee has no problems with the cloning of plant life and animals. It's the cloning of a human being, made in the image of God, that worries him.

There are very good reasons to stop this human cloning train before it crosses into the land of the forbidden, both from a scientific and a Christian perspective. We'll consider these in the next two chapters.

4

Should Humans Be Cloned?

Scientific meetings aren't always the stuffy sessions they're made out to be. At a March 1977 forum of the National Academy of Sciences in Washington, D.C., the subject was cloning. A learned researcher was lecturing when a band of objectors suddenly began chanting, "We shall not be cloned!" Swarming onto the stage, they unfurled a banner proclaiming: "We will create the perfect race—Adolf Hitler." And that was more than twenty years ago.

Scientific debate about cloning is usually a little more orderly but still spirited. Rarely will you find anyone neutral about whether human beings should be cloned, especially with the recent advances in that direction. Molecular biologist Leon Kass once served as executive secretary of the Committee on Life Sciences and Social Policy for the National Academy of Sciences. He told an annual meeting of the Association of American Law

Schools in Chicago that a human being should not be cloned even once. The ability to do so, he said, is not a justifiable reason. Princeton's Dr. Paul Ramsey, the theologian most often quoted by those against cloning, said prospective mishaps in cloning experiments that would inevitably result in the formation and destruction of malformed embryos call for a moral prohibition.

Stanford's Dr. Joshua Lederberg and Joseph Fletcher, of "situation ethics" fame, have been two of the main cheerleaders for human cloning. Lederberg has never advocated doing away with the old-fashioned way of having babies, but he does think that clonal reproduction should have equal opportunity. Fletcher's dictum is that humanity itself must be in control of human evolution. This is typical of the way evolutionary humanists think. Humanity, they believe, is not here by design and special creation but only by evolutionary processes. Since there is no creator, sustainer, and controller of the universe, we must look out for ourselves and work out our own destiny.

Everybody, it seems, has an opinion about cloning. A young humorist named Rodney Cook can't understand why some people are opposed to it. "This could be the answer for society's woes," says Cook in a small Arkansas county paper.

I see in the future that people will have their own personal cloning kits with sales slogans like these: "For those who can never get a date, you need Clone-A-Date! Cindy Crawford can be your escort to the high school dance. Chelsea Clinton can be your companion to the fraternity social. No need to be without a date again!

Also for those couples who feel that kids today are just going to pot, you can "Create-A-Kid"! Make a family completely in your image with memory implants and personalized morals! No more concerns with your child doing drugs or joining gangs, because your child will be you.[1]

All joking aside, cloning is a serious, even frightening prospect for the future. In this chapter I'm going to try to refute the stated advantages of human cloning cited by those who think it would be a good and daring thing to do.

Debunking the "Advantages" of Cloning

Cloning Is a Great Way to Perpetuate Genius

How about a hundred Einsteins, two dozen Picassos, ten or twelve Mozarts, a dozen Shakespeares, and a lifetime supply of Carusos? The fact is that we cannot clone people who are long dead and buried. Even if we exhumed their bodies and searched for live cells, there would be none to find. The way we bury people, the cells of the deceased expire pretty fast.

Actually, the promoters of cloning advocate selecting a few of the "greatest" men and women now living for cloning. But who would select the candidates? If the cloning were to be done by a government or university team of biologists, a committee would probably be named. What criteria would they use? Would Christian faith be a plus or minus? Right now, evolutionary humanists are pretty much in control of our higher educational system, the textbook publishing houses, and the government bureaucracies that fund most of the scientific research. I doubt if they'd like a dozen more Billy Grahams or Jerry Falwells.

Even if an exact chip off the brilliant block of Albert Einstein could be cloned, there is no guarantee that his "twin" would be as smart as the original. Scientists continue to debate the relative powers of nature and nurture in shaping a life. Is heredity or environment more important? Natural intelligence or the motivation to learn? We all know gifted people who choose to dissipate their endowments through laziness, worthless hobbies, drug use, or some other desecration of talent. Furthermore, the clonal "son" or "daughter" of a genius would live in another era. The time and setting in which we live affect the way we respond to challenges.

Also, numerous studies have shown that talent and accomplishments tend to run in families. Children of doctors are more likely to become physicians, for example. Children born and nurtured by parents who value learning are more apt to graduate from college. There are exceptions, of course, but in general this is the way it is. Some studies have shown that the home has more influence on a child's development than today's public school.

Unless adopted, clonal children would not have normal family lives. How would they fare in peer relationships? The "parents" of clones might decide to keep their little group of identicals together. How would it be for three or four identicals to grow up in the same household? Suppose clonal children knew they were the offspring of geniuses. How would they be affected by the social pressure to measure up? Any youth counselor can tell you about the agony some kids endure in trying to meet parental expectations.

There's a related issue to the cloning of genius that I must mention. This is the possibility of a megalomaniac having himself cloned. Or a deranged scientist deciding to clone his beloved despot. If you want a graphic picture of the horrors this might bring about, read a book I cited in chapter 1, Ira Levin's novel *The Boys from Brazil*. The book opens with former Nazi officials being given orders to commit a bizarre series of murders of sixty-five-year-old adoptive fathers. As the plot unfolds,

Nazi hunter Yakov Liebermann follows the trail to Frieda Maloney, a former concentration camp guard, who was hired by an organization to get a job with an adoption agency so she could look at their files. Following instructions, she selected applications of would-be adoptive parents in which the husband was born between 1908 and 1912 and the wife between 1931 and 1935. Each husband had to have a civil service job, and both parents had to be white Christians with a Nordic racial background. At regular intervals, the organization gave her babies to place with these families. She thought they were the illegitimate children of German girls and South American boys and innocently went along with the scheme. Actually, they were the progeny of Hitler, cloned from cells taken from his body in 1943 and preserved for future cloning by the evil Nazi physician, Dr. Josef Mengele. The babies—ninety-four little Hitlers—were conceived in Dr. Mengele's lab and carried to term by Brazilian Indian tribeswomen.

Dr. Mengele planned for them to have parents similar to the Fuehrer's. Hitler's father was a civil servant, a customs officer. He was fifty-two when Adolf was born; his wife, twenty-nine. The father died at sixty-five when Adolf was almost fourteen. In the book, the adoptive fathers were killed systematically about the time of their sixty-fifth birthdays, when their adopted sons were almost fourteen. In a chilling scene near the end of the novel, Mengele meets one of the boys. "You were born from a cell of the greatest man who ever lived!" he shouts. "*Re*born! . . . You're the living duplicate of the greatest man in all history!"[2] I don't want to give the rest of the story away. Read the book, and you'll find the boy's response intriguing.

The idea of cloning genius is not heaven-sent. It comes from another source.

Cloning Can Provide Soldier and Servant Classes of People

Joseph Fletcher once suggested that top soldiers might be cloned to fight soldiers of a tyrannical power. As far out as this sounds, I find it interesting that he admits that despots will use cloning. This isn't science fiction. If one person can be cloned, an army can be so produced. Presumably, the "parents" would all be fine physical specimens with keen minds and quick reflexes to pass on to their clones.

The raw material could be warehoused in special clone banks. Donor inseminators are already using sperm banks. Remember Robert Graham, who died about the time Dolly the sheep was in the news? Several years ago, he set up a special bank in California to house the sperm of Nobel prize winners. Columnist Ellen Goodman says as many as 218 children have already been born "with sperm from those little narcissus bulbs."[3]

Soldiers are not the only special group clonal crusaders have in mind. Some universities are so hard up for football talent that they might even guarantee alumni that funding for cloning projects would result in winning teams. Also, special clones could be produced for other jobs that ordinary humans find distasteful or dangerous. Scientists would be cloning people to be used as slaves and work animals. They would look like ordinary human beings, yet they would be treated as subhuman. They would have no more rights than animals. I can't imagine any humane person backing such a program.

Cloning Can Improve the Genetics of the Human Race

Selective breeding of subjects has been practiced by slave holders from Roman times to the pre–Civil War

American South. They simply ordered strong, intelligent men and women to sleep together, even if they were betrothed or married to other people. Hitler's scientists probably never considered human cloning possible when they set out to purify the "Aryan race." They did it the way previous tyrants had. If the Nazis had known how to clone, they certainly wouldn't have gone to the trouble of bringing in young Aryan men and women for mating.

Cloning will likely bring about a genetic disadvantage for race improvement, depending on how much the procedure is used. The cloning of millions of people will diminish the variability resulting from the heterosexual mixing of genes. Generally, the combining of two different heredities will produce a stronger progeny. For this reason a farmer buys hybrid seeds instead of using his own from a single stock. Another problem will become clearer as I move forward. Everyone possesses harmful genes that result from mutation. Cloning will keep these genes in circulation. Couldn't persons with problem genes be identified and classified "4-F," like draft rejects were in World War II? Probably not, unless some physical evidence of genetic disease is present. That type of person wouldn't be considered by doctors in a despotic nation.

Dr. Lederberg, at Stanford University, conceded that wholesale clonal reproduction could lead into an "evolutionary cul-de-sac" with no chance of genetic improvement. A quarter century ago he proposed "tempered clonality," which would allow for both cloning and heterosexual reproduction. He thought then that selected cloning could improve a race by adding "better" people. This would require a government regulatory agency with two major responsibilities: (1) to select the candidates for cloning, and (2) to keep the clonal people from mating with other clonal people or the choice offspring of heterosexual mating. Allowing clonal people to mate would destroy any racial "benefits" to be gained by the selective use of cloning.

With human nature as it is, I don't see the latter as possible. A clonal person would have the same urges as we all do, and he or she would find a partner. The "crime" would be found out when the child was born. Would the violators be punished for breaking the law of the clones? Would the innocent child be killed?

Cloning Can Prevent Genetic Disease in a Selected Posterity

This argument also sounds good on the surface. The genetic engineers would simply pick out the best "stock" for cloning. Genetic disease caused by the mixing of certain heredities would be avoided. However, this is not as fail-safe as it might seem. Clonal people would still be susceptible to mutations or mistakes in the replication of cells. They would also be affected by environmental influences that may bring out previously unknown genetic disease. One of the many genetic diseases that can remain undetected until activated by environment and lifestyle is Herniske-Korsakoff syndrome. An enzyme called transketolase, which regulates or filters vitamin B_1 to the brain, is missing from the cells. People with the defect don't get enough B_1 to prevent the disorder from spreading. Without B_1 supplements, victims go hopelessly insane.

Cloning will merely perpetuate such genetic defects. It offers no panacea for the elimination of genetic disease.

Clones Can Exchange Body Parts and Experience Enhanced Social Communion

This is one of Dr. Lederberg's key planks in his platform for cloning. Since all clonal persons are identical twins, a number of donors could be available for organ transplants. As identical twins, they would share a closeness and understanding that would facilitate harmony and peace, enabling them to work together cohesively. This assumes that the cloned individuals would be willing to make the sacrifices required in donating organs. It assumes they would be free from self-centeredness, jealousy, greed, and other failings of ordinary men and women. Utopian experiments have always failed to create perfection among ordinary people. I cannot believe that members of a clone would live together in sweet communion. Evolutionary humanists ignore the sin factor that inevitably spoils man-made Edens. "The crude evidence in human experience," said Dr. Paul Ramsey, "does not lend unequivocal support to the expectation that 'intimate communication' would be increased." To the contrary, Ramsey suggested that "animosity in personal relations might be heightened."[4]

Ramsey further indicated that the struggle for selfhood and identity in a clone could be intense. "Growing up as a twin is difficult enough. . . . Who then would want to be the son or daughter of his twin? To mix the parental and the twin relation might well be psychologically disastrous for the young."[5] Personal and psychological independence, Ramsey projected, might be impossible to achieve for children who are the exact copy of their one "parent."

Cloning Can Provide a Genotype of One's Spouse, Living or Dead, of a Deceased Parent, or of Some Other Departed Loved One

Advocates of cloning suggest that couples unable to have children together might prefer a cloned child to one by artificial insemination, since a third person would not be involved. Here, too, there could be an identity problem, unless the child was never told of his or her true parentage. Even so, the amazing likeness between the child and one of the parents would be obvious when the child looked in the mirror. Cloning to preserve the genotype of a departed loved one will require that the necessary cells be removed before death. Biologists can already keep cells alive in a laboratory culture for an indefinite time. If and when human cloning is accomplished, such "restoration" of the dead will be possible. Dead men can even now sire children through artificial insemination, using sperm donated and frozen while they were alive.

Such a possibility raises a hard question. Whose child will the offspring be? The scientist who does the cloning, the relative who pays for the act, or the deceased donor?

Cloning Can Provide a Form of Immortality for Donors

If so, it will be only for those who can afford to pay for the cloning. But let's suppose cloning for this purpose is made available to whoever desires it. We've all known parents who seek to relive their youth through a son or daughter. This can create serious psychological problems in the offspring. No youth should be burdened to live the life of a selfish parent. Furthermore, there is absolutely no evidence that human consciousness can be transferred to a clonal descendant. The genes will be substantially the same, but the memory and thought patterns of the parent will not be reimplanted.

Cloning Can Determine the Sex of Future Children

Aristotle recommended that Greeks have intercourse in the north wind if they wanted males and in the south wind if they desired females. Women in the Middle Ages who wanted boys were told to drink a mixture of strong wine and lion's blood. Men in some European countries wore their boots to bed when they wanted to conceive a boy. Medical science today can give concrete help to improve the chances of producing the desired sex. Dr. Landrum Shettles is counted as a pioneer in this area. It was known for many years that if the sperm that fertilizes the egg carries an X chromosome, the baby will be a girl; if a Y, the child will be a boy. The problem was that no one had been able to distinguish between androsperm (male sperm) and gynosperm (female sperm). Shettles examined some living sperm under a phase-contrast microscope that revealed details ordinary microscopes had missed. He concluded that the male sperm were round-headed and the female oval-shaped and that the males outnumbered the females in the ejaculate. Further experiments led him to believe that the larger number of androsperm was nature's way of compensating for their low resistance to acid secretions in the female vagina. He believed if the sperm could break through this acid barrier, they would have a better chance than female sperm of surviving in the alkaline environment within the cervix and uterus. From his observations, he compiled a list of procedures that he advised couples to follow in sex selection of children. In one case study, nineteen of twenty-one couples who wanted girls were successful. In another test, twenty-three of twenty-six couples wanting boys got their hearts' desire.

Then, while researching scientific and historic literature on sex selection, Shettles found that Orthodox Jews produce more male offspring than does the general population. In the Jewish Talmud he found that rabbis had proposed one of the procedures he had devised in his lab. Dr. Shettles' sex-selection techniques are not accepted by all doctors. He never guaranteed that the techniques would always work, only that they would increase chances of conceiving a child of the desired sex.

The greatest strides in sex selection are now being made with animals. Scientists at the University of Missouri worked with the U.S. Department of Agriculture (USDA), using gender-separated semen to fertilize sow eggs in test tubes, which were later implanted in surrogate mother sows. The gender separation of semen had been developed by the USDA several years before. The process separated semen into two components: one contained sperm carrying the X, or female, chromosome and the other held the Y, or male, chromosome. The sorting was made possible because the X chromosome is larger than the Y chromosome. When the piglets were born, the scientists found they had achieved 85 percent success in predicting the gender.

Scientists say the successful experiment can improve hog production and engineer hogs genetically to provide hearts, kidneys, and other organs for transplant purposes.[6] This is very recent, and I have seen no predictions that the process could be applied to humans. I can only assume that this is possible.

The prenatal test known as amniocentesis can now be used to identify the sex of a child before birth. The test is designed to identify certain genetic defects; however, doctors are faced with more and more parents demanding the test for purposes of sex selection. If the child is not of the sex desired, the parents request an abortion. Such callousness is a mockery of God's gift of life. More than that, abortion for this purpose is, in my view, first-degree murder. It lowers biological parents below the level of animals who would not reject their own flesh and blood.

With cloning there is no need to follow complicated procedures in intercourse or to have the amniocentesis test. The cloning of a man will always produce a male and of a woman will always result in a female.

Cloning Can Increase Scientific Knowledge about Human Reproduction

Hitler's scientists increased their knowledge by studying human brains taken from the skulls of undesirables killed in pruning people from the Aryan race. This, of course, did not justify the murder of these innocent people. Unlawful and inhumane experiments on other human beings have occurred, even in the United States. I don't doubt that some American scientists have worked with aborted, live human fetuses. Their goal may be to learn more about genetic disease, but the work is still wrong—shockingly, appallingly wrong—and a gross violation of the rights of the unborn. I have already said that a cloned offspring, prenatal or postnatal, will be a person. He or she should not be treated like a laboratory animal.

Conclusion

These then are the nine most often mentioned scientific justifications for cloning humans. Cloning advocates differ over which ones they consider most important. Some scientists who favor cloning reject certain of the more trivial possible uses, such as cloning to determine the sex of a child. A few scientists mention more fanciful reasons, such as cloning legless astronauts to take long journeys into space in cramped space capsules. This is more in the realm of science fiction than the cloning I discuss in this book. I am restricting my material to what I believe is possible within the next few years.

5

Cloning Is Not God's Way

In the previous chapter, I sought to present some of the loopholes in arguments for human cloning. Now I want to speak from a Christian perspective. The Christian worldview is markedly different from that of the secularist, who does not accept the idea of God or a purposeful, orderly universe. Secularists may call themselves humanists; however, if they are true to their philosophy, they will be more like mechanists, who see humans in terms of function rather than of being. They will hold society or the common good in greater reverence than the individual. They will make individual aspirations and goals subservient to the good of the species—as they see it.

Christians, however, begin with God, not people. We believe God has revealed his will for humanity in the Bible, with the foundations laid in Genesis. All that we see in nature and discover through human experience and scientific experimentation merely furthers our understanding of God's revelation in the Bible.

God and Science

I'll give you a couple of examples. The Hebrews of Moses' day certainly didn't understand the laws of genetics. They had never studied cells under an electron microscope, and they no more understood the relationship between recessive and dominant genes than they did the functions of electrons and protons in the atom. But God told them they were not to marry close blood relatives (Lev. 18:6–16). He didn't tell them why from a scientific standpoint, yet we know that the levitical law against incest makes good genetic sense. One who marries a close blood relative risks "stacking" bad genes that may result in retarded or diseased offspring.

The prohibition against adultery was also plain enough to God's ancient people. They didn't know, as we do, that the seventh commandment makes good sociological and psychological sense. They didn't have the statistical techniques and data processing equipment we can use today to prove the wisdom of this scriptural prohibition. These and other laws relating to marriage and family were given thousands of years ago. Yet they're as up-to-date now as when they were given. God created human beings with the freedom to choose between right and wrong. He presented us with options and spelled out the consequences of the wrong choices.

If human cloning is possible within the biological mechanisms God has created—and I think it is—then short of the end of the world, God will not stop it from happening. If a human cell nucleus with forty-six chromosomes can be inserted into an enucleated egg cell and be chemically triggered to start dividing and growing, then the resulting embryo will develop and grow into a fetus just as in ordinary reproduction. This will not be *creating* new human life but simply *reproducing* life in a novel way. If it happens, I will not stop believing in God or the Bible. I will view cloning as just another attempt by humanity—albeit one with potentially grave consequences—to circumvent the plan of God for the perpetuation of the human race. Human cloning will not, however, be wrong because it is a new achievement in science. We Christians have too often been guilty of condemning a scientific development simply because it is new. Take this example from early America. As a young man preparing for the Congregational ministry in Braintree, Massachusetts, John Adams became upset over the condemnation of the smallpox vaccination. Leading clergymen were calling smallpox a divine judgment for wickedness. It was wrong, they said, to interfere with the carrying out of God's judgment. Because of such absurd dogmatism and other wrangles within his church, Adams switched to the law profession and later succeeded George Washington as second president of the United States.

I want to say it loud and clear—neither genetics nor any other division of science is inherently an enemy of the Bible and Christian faith. Science is neutral, amoral. It can be used for good or evil. Indeed, the mandate for honorable scientific research is found in the Genesis command to "subdue" the creation. Twenty years before the cloning of mammals, Donald MacKay, a highly renowned English brain physiologist said:

The key to the whole problem of the relation of science to the Christian faith is that God, and God's activity, come in not only as extras here and there but everywhere. If God is active in any part of the physical world, he is in all. If divine activity means anything, then all the events of what we call the physical world are dependent on that activity.[1]

MacKay defined the role of scientists as "mapmakers in God's world. . . . It is not just that we have permis-

69

sion, but that we have some obligation to get out there and map it so that the people who come after us can use the map. Mapmaking is wrong," clarified MacKay when it "involves an infringement of the moral law. One should not become an adulterer in order to study the psychology of adultery. Apart from moral considerations, I know of no biblical guidance that prohibits any particular area of research."[2]

MacKay added, "Cloning [humans], as such, does not seem to me to transgress any biblical injunction."[3] The English scientist, who has spoken to many Christian groups, said clonal offspring will be no different, in terms of human dignity, from identical twins. I disagree. There will be a great deal of difference. The birth of identical twins will continue to be a pretty rare occurrence. How many do you know? Once cloning is perfected, we may be faced with thousands, even millions of such people. Given the right circumstances, a thousand identical clonal beings could inhabit one city. Sociologically, that is mind-boggling.

However, MacKay qualified his endorsement of cloning by saying it might not be desirable to replenish the earth by reducing the variety of genetic types, which would happen in cloning. "To mess about with such a delicately balanced system without good reason could be disastrously irresponsible. It would also be very difficult to ensure proper psychological development in cloned children deprived of a normal family background."[4] Here, in my judgment, MacKay comes close to the crucial biblical objection against cloning: the lack of a normal family arrangement.

The Family: God's Design

God knew that children would need two parents of complementary yet different sexual identities. "In the image of God He created him; male and female He created them. And God blessed them; and God said to them, 'Be fruitful and multiply'" (Gen. 1:27–28). Clearly God intended society to be built on the two-parent family. We're now suffering from an epidemic of divorces. Single-parent families comprise a large segment of American society. Dr. Wade F. Horn, a well-known clinical psychologist, is director of the National Fatherhood Initiative. In a 1997 seminar at Hillsdale College's Shavano Institute for National Leadership, Horn noted that in 1960 the total number of fatherless families in the United States was less than eight million. Today, that total has risen to nearly twenty-four million. Nearly four out of ten children in America are being raised in homes without their fathers. "How did this happen?" he asked. "Why are so many of our nation's children growing up without a full-time father? It is because our culture has accepted the idea that fathers are superfluous—in other words, they are not necessary in the human family."[5]

In millions of U.S. households, women are shouldering the load alone, a dramatic increase in the past decade. In many inner-city public housing projects, more than 90 percent of the families are without a father. The crime rate in these projects is staggering. Willful, sinful rebellion is the root factor, but this is abetted by poverty, unemployment, and the absence of paternal guidance and masculine role models. I don't doubt that the single largest cause of the upsurge in crime is the breakdown of the family structure. Hear what Dr. Leanor Turoff Glueck, a criminologist researcher at Harvard University, has said:

> Every child has a right to parents who love each other, who want him and love him, who are deeply concerned for his welfare, who will give him supervision so he will not be so completely left to his own resources that he

71

will make serious mistakes. . . . Unless and until there is a concerted effort to preserve good families and reconstruct those that are not good, we are going to see crime penetrate every social level.[6]

God could have created humanity as a single sex. He could have created us as two or more sexes, each under one parent, to be perpetuated by cloning. But he didn't create just Adam or just Eve, or Adam as the king of one kingdom and Eve as the queen of another. He made Adam and Eve partners in reproduction and in the care of children, setting the pattern for future generations. "For this cause a man . . . shall cleave to his wife; and they shall become one flesh" (Gen. 2:24).

In God's marriage mathematics, one plus one usually equals one, although it can also produce two, three, or more (multiple births). Still, each child has his or her own identity, even when he or she is an "identical" twin. The two parents still see their expression of love in each individual child. Most families have a mixture of males and females. The males are outnumbered in my family three to two. Even in units in which the children are all girls or all boys, there is still one parent of the opposite sex, unless death or divorce has intervened.

Clonal persons must necessarily be all of one sex. A female clone can only have a female "parent"; a male clone only a "parent" of his sex. This may mean, because of the difficulties inherent in a single-sex family, that clonal children will not be raised by their "parent." It is likely that they will be raised in nurseries, and sociological studies have repeatedly shown that children raised in nursery settings are more prone to develop physical and mental problems than are children raised in families.

One of the classic experiments in child development was conducted many years ago by Dr. René Spitz of New York. He surveyed children in two similar institutions, which differed only in the type of personal care offered. "Nursery" babies were cared for by their mothers. "Foundlinghome" infants were given to the care of a few overworked nursing personnel. A guide called Development Quotient (DQ) was used to compare the development of the two groups. This included control of bodily functions, muscle coordination, social relations, and general intelligence. "Nursery" babies started with a DQ of 101.5 and rose in one year to 105. "Foundlinghome" babies began with 124 DQ and declined in two years to 45. In two years 37 percent of "foundlinghome" children died, but in five years not a single "nursery" child was lost. Clearly, all children, including clonal offspring, need attentive, involved, loving care. Without it, they die.

We need to consider the importance of role models and relationships in a family. Two parents are needed for the development of good sexual identities in children. Many, if not most, in the homosexual population come from homes devoid of healthy role relationships with parents. If healthy homes are absent, it is entirely possible that large numbers of clonal children will turn toward homosexuality or other sexual dysfunctions.

You've likely heard of the yin and yang principle. Yang is the masculine or active force, yin the feminine or passive force. Taoists believe the interplay of these forces occurs in every activity of the universe. At one time in one place yang may be stronger; at another time and location yin may dominate. This is a beautiful illustration of the give-and-take in the marriage relationship, reproduction, and the family interactions that support the continuum of life. Human cloning, if done on a large scale, will probably upset that God-ordained balance of gender. And we will all be the losers for it.

I agree with Paul Ramsey that cloners will most likely be no more than technicians in animal husbandry. In their attempt to improve humanity, they will lower humankind. Ramsey warned:

> To put radically asunder what nature and nature's God joined together in parenthood when he made love procreative, to disregard the foundation of the covenant of marriage and the covenant of parenthood in the reality that makes for at least minimally loving procreation, to attempt to soar so high above an eminently human parenthood, is inevitably to fall far below—into a vast technological alienation of man. Limitless dominion over procreation means the boundless servility of manwomanhood. The conquest of evolution by setting sexual love and procreation radically asunder entails depersonalization in the extreme. The entire rationalization of procreation—its replacement by replication—can only mean the abolition of man's embodied personhood.[7]

The Sanctity of God-Created Life

Another major objection to human cloning from a Christian and humane point of view relates to the destruction of clonal life in experimentation. Such wanton destruction of unborn human beings, whether in embryo or fetus form, must be opposed. This is not just a conservative Christian position. It is also strongly held by many other groups in our society who believe that the right of the unborn to live should take precedence over the curiosity of scientists to experiment.

Some time ago, when the possibility of human cloning was just beginning to be featured in popular media, Dr. Leon Kass responded to a column promoting cloning by Dr. Joshua Lederberg in the *Washington Post*. In a letter to the editor, Kass asked what would happen to mishaps in cloning. "Who will or should care for 'it' and what rights will 'it' have?" he wanted to know. "Will the programmed reproduction of man not in fact dehumanize him?" Kass further wondered if it made "sense to say that each person has a right not to be deliberately denied a unique genotype. Is one inherently injured by having been made the copy of another human being, regardless of which human being?"[8] Long a thorn in the side of scientists who live by consequentialist ethics, Kass held that even to attempt cloning is unethical because it will "place in jeopardy the clonal offspring, the permission of which to do so we cannot obtain."[9]

The decline of reverence toward fetal life since the Supreme Court voted to allow abortion on demand is already dismaying. When I was in western Canada on a lecture tour some years ago, I discovered that this malaise of the spirit is not confined within our borders. One night I was interviewed by a reporter in Saskatoon, Saskatchewan. He mentioned that his sister was a neonatal nurse, one who works with premature infants and babies with birth defects. "How does she feel about abortion?" I asked. "Oh, she's dead-set against it," he replied. "What really bothers her is that in one wing of the hospital they're killing babies older than the ones she's helping to save in her wing."

A million and a half babies will be aborted this year in the United States alone—more than fifty million worldwide. Population explosion or not, this is nothing short of mass murder of those who cannot protect themselves. Cloning, accompanied by the destruction of laboratory mistakes, will only add to the callousness that threatens us all. I am totally against this scheme to propagate children by asexual reproduction. I don't see an embryo or fetus as just a pulsating blob of tissue. This precious life, whether in a woman's womb or in a laboratory culture, is the fruit of God's own creative processes, a human in miniature, and at the very least a life in the making. The right to live should be preeminent.

The Science Committee of the U.S. House of Representatives recently passed by voice vote the Human Cloning Research Prohibition Act, H.R. 922. The bill, which also must be approved by the Commerce Committee before going to the House floor, prohibits federal funds for research involving the use of a human cell other than that of a sperm or egg to produce an embryo. It calls for recommendations within five years from the National Research Council on whether changes should be made in the law. Unfortunately, at this writing, the bill has not cleared the legislative hurdles, enabling it to become law.

The bill is flawed on three counts, according to Ben Mitchell, a biomedical issues consultant for the Southern Baptist Ethics and Religious Liberty Commission. Mitchell, in a letter faxed to the Commerce Committee, said the legislation falls short, first, because it bans only federal funding of human cloning research instead of a prohibition on all such research. (The Science Committee, however, has jurisdiction only in funding. Another bill, HR923, would prohibit human cloning research in the private sector.) Second, it fails to define "embryo" in a way that would cover fertilization to birth. And third, it allows Congress to revisit the issue within five years, even though the "moral status of unborn life" will not change during that time. [10]

Mitchell is right on target with his pro-life concerns. Human life is too precious for experimentation by would-be cloners, whether they are mere graduate students or Nobel prize winners. In the blind rejection of the Creator's plan, they may honestly feel that cloning will benefit the human race. No matter. They are not wiser than God. They should not be allowed to carry on their experiments when human life is at stake. We should rise up and say so now before C-Day—before a team of scientists stands up before the world and grandly announces, "We have cloned a human being."

6

Test-Tube Babies

While a biology professor was away from his lab, his students decided to have some fun. They glued together an assortment of body parts from various bugs—the legs of a grasshopper, the wings of a butterfly, the torso of a black beetle, the head of a caterpillar—and placed the creation on the teacher's desk. When the professor returned, a student pointed at the desk and asked, "Prof, how would you classify this, uh, specimen?" The teacher picked it up gingerly and turned it over, pretending to inspect it closely. Then he announced with great solemnity, "Ladies and gentlemen, this is a humbug!"

You may still believe that cloning a human being is all humbug—foolish speculation by a few scientists trying to grab public attention. But you can't say that about the so-called test-tube baby. Baby Louise Brown was born by cesarean section at Oldham General Hospital in northern England, 18 July 1978. That was twenty years ago! She is the first child known to have been conceived in a laboratory culture dish. Gynecologist Patrick Steptoe, then a staff doctor in Britain's National Health Service, and Cambridge University physiologist Robert Edwards share the honors of bringing about this amazing scientific achievement. The announcement about the test-tube baby kicked off a media circus. *The London Daily Mail* bought pictures and exclusive rights to the Louise Brown story from her parents. The publicity extravaganza took the family to Disney World in the United States and to Japan, where crowds gazed at the blonde, blue-eyed baby in amazement. Her parents, truck driver Gilbert Brown and his wife, Lesley, received enough money to move from their small house in the British port city of Bristol to a large country home.

No birth of a princess ever excited so much attention. The event was headlined around the world. Many reporters compared the scientific breakthrough to the splitting of the atom and predicted it would have far greater consequences. That, I repeat, was twenty years ago. The next greatest media excitement burst forth in 1997, when the cloning of Dolly was trumpeted from Scotland. Can we even imagine the media earthquakes that will accompany the cloning of a human being?

Biologists were not taken by surprise when Baby Louise was born, having been conceived in a test tube. However, they did not all agree that a test-tube baby was a good thing. Dr. James Watson, who had shared a Nobel prize with two colleagues for unraveling the structure of DNA, the chemical blueprint of life, had warned a U.S. congressional subcommittee years before that such a birth would occur, predicting explosions, politically and morally, all over the world. That hardly happened. After the headlines and a flurry of television specials, the excitement cooled. Most doctors wondered what all the fanfare had been about in the first place. They saw no moral problem as long as it was a husband's sperm and a wife's egg that had been mated in the culture dish. Human life had not been "created" in the laboratory. Steptoe and Edwards had only built a detour around the wife's blocked oviducts for the sperm and egg to achieve fertilization. This, they said, was a godsend to the 1 percent of women who cannot conceive because of this problem.

These assessments, however, fell short of the mark. The test-tube baby did not by herself usher us into a "brave new world" of mechanical reproduction. When considered in the larger context of the biorevolution, though, it portended far-reaching implications. For one thing, it brought human cloning a giant step closer.

To understand what Steptoe and Edwards accomplished and some effects of their triumph, we should review the early stages of the human reproductive process.

The Reproductive Process

The female reproductive organs are shaped something like a T, with the small twin ovaries located underneath the curving arms of the oviducts. Millions of sperm from the male enter the uterus and swim "up" to the point where the oviducts branch off at the top of the T. If it is the right time in the menstrual cycle, an egg has broken free from an ovary and meets the sperm in one of the tubes. Conception occurs when the egg and one successful sperm cell join. The fertilized egg, a new human life called a zygote, travels down this tube and burrows itself in the lining of the uterus, where a nesting place has already been prepared. Snug and nourished by the mother's bloodstream (barring damage by accident or disease), the embryo will develop into a fetus to be born at the appropriate time. This is the normal process.

When blockage of the oviducts occurs, however, it is like having both lanes of a divided tunnel barricaded. The crucial connection between sperm and egg cannot be made in either lane. This was the condition that prevented Lesley Brown from conceiving the normal way. What Steptoe and Edwards did was to remove the sperm and egg and enact the crucial fertilization in the laboratory.

Steps along the Way to In Vitro Fertilization

History shows us that often one scientific achievement builds on a previous work. Steptoe and Edwards were no exceptions. In the 1940s Dr. John Rock of Harvard, "father" of the Pill, simulated the chemical environment of a woman's oviducts in a laboratory culture dish. He succeeded in fertilizing eggs (taken from female cancer patients) with sperm and brought several embryos to a multicell stage. In 1950 Dr. Landrum Shettles kept a laboratory-conceived embryo alive for six days outside the womb. Eleven years later, Dr. Daniele Petrucci of the University of Bologna kept an embryo alive for twenty days, until it became deformed. The following year, Petrucci nourished an embryo for fifty-seven days, during which the Italian scientist recorded fetal heartbeats on an EKG machine. Communist Chinese newspapers called Petrucci's experiments happy news for women, providing hope for the development of methods of reproduction that would not burden working women with childbirth. In Italy, the news of what Petrucci had done caused an uproar. Blasted with criticism from the Vatican, Petrucci moved to the Soviet Union and joined a team of Russian scientists at work on an artificial womb.

In 1971 at Columbia-Presbyterian Hospital in New York City, Dr. Landrum Shettles fertilized an egg from one woman, grew it to the sixteen-cell blastocyst stage, and implanted it into the uterus of a second woman scheduled for a hysterectomy. The embryo grew to several hundred cells and was then removed during surgery. A Florida dentist and his wife heard about Shettles's work and asked if he would help them have a baby. Shettles saw the woman, Doris Del Zio, as an ideal candidate. She was healthy, young, and her oviducts were blocked. In cooperation with Dr. William J. Sweeney, Shettles fertilized Mrs. Del Zio's egg with her husband's sperm and achieved conception in the lab. Two days before the embryo was due to be implanted into the woman's womb, the developing life was allegedly confiscated by Dr. Raymond Vande Wiele, Shettles's superior at Columbia. Vande Wiele later told David Rorvik that Shettles had never obtained clearance for the "illegal" and "immoral" project. Shettles held that permission was not necessary, since Mrs. Del Zio was Dr. Sweeney's patient and the actual implant was to have been done at another hospital. Shettles added that Vande Wiele "became emotional" and ordered him out of the hospital. The Del Zios sued for $1.5 million, contending that the hospital and Vande Wiele had deprived them of their property and that substantial damage had been done to Mrs. Del Zio's psyche. A jury agreed but awarded only fifty thousand dollars. The legal hassle between the two contending doctors never reached court. Shettles resigned "voluntarily" from the hospital staff.[1]

Meanwhile, English scientists were experiencing no legal difficulties in attempts to produce a test-tube baby. Dr. Douglas Bevis made the surprise announcement that he had presided over the birth of three test-tube babies by the IVF method. Bevis freely admitted previous failures resulting in the death of embryos. "So many have been attempted," said Bevis, "that by the law of averages some have come through."[2] Bevis's claim was not accepted by his colleagues because he produced no proof. But newspaper stories of his and Shettles's work did arouse opposition from many church leaders and some scientists. Some of the churchmen quoted German theologian Dietrich Bonhoeffer, a martyr of Nazi Germany during World War II, who argued that an embryo's existence is itself evidence of God's intention to create a human being; therefore, the embryo's right to life is divinely bestowed, and any deliberate deprivation of it is nothing short of murder.[3] Criticism over destruction of embryos in test-tube baby experiments resulted in the U.S. government banning new grants for in vitro fertilization. Denied federal money, IVF work in the United States virtually came to a halt.

Steptoe and Edwards—and Beyond

The English doctors Steptoe and Edwards had continued to push ahead. When the Browns came to them for help in 1977, they were ready. In the successful experiment, Steptoe, the surgeon, made a small incision just below Lesley Brown's navel. He inserted tiny forceps, grasped the ovaries, and extracted about fifteen tiny eggs. He then transferred these eggs to a culture dish containing a special mix of hormones and other nutrients that he had concocted to simulate the womb environment. Edwards later removed and washed the eggs in a chemical solution before mixing them with fifty thousand or so of Mr. Brown's sperm cells. Periodically, Edwards and Steptoe examined the eggs and sperm under a microscope for evidence of fertilization. When this occurred, they transferred the new embryo to still another dish. Within four days it grew to thirty-two cells. Mrs. Brown was brought back to the operating room, where Steptoe used microscopic instruments to insert the embryo into her womb. Slightly less than nine months later, Baby Louise Brown was born.

Steptoe came to the United States and presented a full account of the sensational birthing to the American Fertility Society at its meeting in San Francisco. He revealed that he and Edwards had made thirty efforts to implant embryos before succeeding. The failures did not appear to bother the twelve hundred doctors and biomedical specialists present; they gave Steptoe a thunderous ovation. Steptoe and Edwards's success stirred U.S. scientists to ask that the ban on federal grants for in vitro research be lifted. The Ethics Advisory Board of the Department of Health, Education, and Welfare held hearings in eleven cities. Those who favored abortion on demand tended to support test-tube baby research. They generally took the position of Joseph Fletcher, that "a fetus is not a moral or personal being since it lacks freedom, self-determination, rationality, the ability to choose either means or ends, and knowledge of its circumstances."[4] Those who wanted the spigot of federal money to remain turned off sought to draw attention to the deaths of the unsuccessful embryos. Princeton's Paul Ramsey also feared that physical or psychological damage could be inflicted on IVF children.

Test-tube baby research continued abroad. A second IVF baby was born in India and a third in Scotland. The technique involved in the conception and implantation of the baby born in India differed from the Steptoe-Edwards method in two ways. First, several eggs were fertilized for implantation, giving better chances of success if a number of embryos were implanted. Second, the fertilized eggs were kept frozen until implantation, which was performed at the time in the mother's menstrual cycle when chances of acceptance were considered optimum.

Few U.S. doctors and scientists seemed alarmed by this. They argued that laboratory conception and implantation of an embryo was no different in principle from taking fertility drugs or having a baby by cesarean section. The question of whether we should interfere with the natural processes of human reproduction, they noted, had already been decided. Norfolk General Hospital had already applied for and been granted permission by the Virginia health commissioner to establish the first test-tube baby laboratory in the United States. Several other U.S. hospitals were planning to open test-tube baby centers as soon as they could obtain official clearance.

The Ethical Ramifications of IVF

Biological scientists generally saw then—and see now—in vitro fertilization as another step in the effort to improve the quality of human life. Laboratory conception and embryo growth, they believed, would enable researchers to study genetic defects more closely. More might be learned about Down's syndrome and other well-known genetic abnormalities. It might be possible, they said, to detect abnormalities by extracting a few cells from an early embryo for study of defects. The embryo could be frozen while the cells were multiplying in a culture and being checked. If the cells proved healthy, the embryo could then be thawed and implanted within the uterus. If the embryo proved defective, it might still be possible to correct malformed or malfunctioning genes. If the disease was deemed incurable, then the embryo could be used for experimental purposes, for example, testing to see how much radiation the cells could endure or determining what drugs were harmful and in what dosage.

Pro-life people then and now say that this is no different from ordinary abortion. Researchers reply that a large percentage of diseased embryos spontaneously abort anyway. Pro-lifers respond that this cannot be determined until the embryo is given a chance to fight for life. Those who believe that unborn human beings have rights also object to this practice on the basis that it is unlawful to experiment on live members of our species. It cannot be justified, they say, on the basis of potential benefit for other people or the human race in general.

It is now unlawful to experiment on live human beings without consent. Before World War II, this was only a consensus in many countries. It was felt that medical science would abide by the unwritten rules. Then the horrors of Nazi concentration camps and experimental hospitals were revealed. The world learned that Nazi medical personnel had been instructed to go through their wards and select the least fit to use for experiments. Some hospital workers picked out known troublemakers and other patients they disliked. Nazi researchers did not care who was chosen, as long as they got their daily supply of fresh human organs. Disclosures of what happened in Nazi Germany resulted in the Nuremberg Code of 1949, the first code of medical ethics for research on humans. This was adopted by the World Medical Association at the first 1964 meeting in Helsinki, Finland. The code called for:

- Experimentation to "conform to the moral and scientific principles that justify medical research" and to be "based on laboratory and animal experiments or other scientifically established facts."
- "Every clinical research project" to "be preceded by careful assessment of the inherent risks in comparison to foreseeable benefits to the subject or to others."
- The doctor "to remain the protector of the life and health of that person on whom clinical research is being carried out."

- The doctor to explain "the nature, the purpose, and the risk of clinical research" to the subject.
- The doctor to "obtain . . . in writing the consent of the subject."[5]

By 1964 the Nazi atrocities were only a bad memory for most people. Trust in the benevolence of doctors continued to climb. Then a U.S. researcher in anesthesiology, Henry Beecher, unveiled a bombshell report, charging that as many as 12 percent of the experiments being conducted in the United States by qualified medical researchers were unethical. Beecher noted that twenty-three charity patients had died after being treated with experimental drugs in an effort to find a better treatment for typhoid fever. Had they received proper care, Beecher said, they would have been expected to survive. He mentioned another case in which a researcher, identified only as Dr. S., injected live cancer cells into twenty-two human subjects, telling them only that they would be receiving "some cells." Dr. S. was later brought before the New York State Board of Regents, the licensing agency for physicians in New York.

An associate of the accused doctor said they had not told the patients the injected cells were cancerous because "we did not wish to stir up any unnecessary anxiety, disturbances, or phobias." A reporter cornered Dr. S. afterward and asked if he would accept an injection of cancer cells into his own bloodstream. He would not, because "there are relatively few trained cancer researchers, and it seems stupid to take even that little risk." In his incriminating report, Beecher said Dr. S.'s research colleagues had not disapproved of his work; indeed the American Association for Cancer Research elected him as its vice president in 1968 and as its president in 1969.[6]

More recently the story of the frightful experiments on African American men with syphilis has been uncovered. It happened at the Tuskegee Institute in Alabama under a government program many decades ago. It is sadly ironic that an institute founded by the great Booker T. Washington should have exploited the trust of the participants. These and other examples of unethical human experimentation are included in *Ethical Decisions in Medicine* by Dr. Howard Brody, a standard text in medical ethics used for many years by some medical schools. Three ethical theories are set forth in this text:

1. Consequentialist ethics, "in which an action is judged to be morally right or wrong by judging the consequence of the action."
2. Utilitarian ethics, which states that the "ultimate principle against which consequences are to be judged is the general happiness of all people concerned, or the greatest net balance of good over evil."
3. Deontological ethics, which insists "that there are rules or principles of action which have moral validity independent of the consequences of individual actions, and that one must act in accordance with these rules or principles." The latter theory, according to author Brody, is usually held by persons with strong religious beliefs.

Brody clearly lines up with consequentialist ethics and suggests that medical practitioners and researchers consider the "risk-benefit" ratio. "The potential benefits to be gained, both by the individual and by society as a whole, must be weighed against the risks of untoward side effects to the individual," he maintains.[7] Consequentialist ethics asks the researcher or his employer to judge whether a specific experiment is right or wrong. I'll admit that we all make decisions every day based on what we think is best. There are simply not enough clear guidelines to cover everything in our modern technological world. But should scientific researchers or anyone else have the authority to decide who will live and who will die through experiments on unborn human life?

This problem, of course, relates to an issue we've already raised: Is an embryo a person? Does it have a right to life? If we answer yes, then there is the matter of consent. Obviously, the embryo's consent cannot be obtained. Then who should consent? The parents? The researcher? Is the good of the human race—a cure for cancer, for example—sufficient reason to give consent? Ramsey thought the health of the individual patient, prenatal or postnatal, must be put before any potential benefits for that "non-patient" (Ramsey's term), the human race. This seems to be a reasonable guideline.

There's another implication of in vitro fertilization we should consider. This is the possibility of a host mother bearing a child in place of the child's biological mother. The host mother concept was demonstrated in animals long before the first test-tube human baby was born. Back in 1962 English scientists extracted two fertilized eggs from sheep, tucked them in the warm oviduct of a live rabbit, and shipped the hare to South Africa. Scientists there removed the eggs and implanted them in two ewes that gave birth to healthy lambs. Here three "mothers" were involved, a biological mother, a transfer mother, and a birthing mother. Scientists in the United States soon did their English colleagues one better. They gave a prize cow fertility hormones to produce a large supply of ripe eggs. They fertilized these eggs with sperm cells from a select bull, then took the resulting embryos and implanted them into the uteri of ordinary cows. Their reward was a whole herd of calves with superior heredity.

The same could be done with a "prize" woman. Give her fertility hormones. "Breed" her to a prize man. Extract the harvest of fertilized eggs. Implant the fertilized eggs in the uteri of a number of healthy women. The result will be a "herd" of "superior" people. Terms such as *breed, prize,* and *herd* fit perfectly into the context of this sort of "animal farm." With a totalitarian government on the order of Nazism or Communism and a few unscrupulous scientists, a whole chain of state-run farms could soon be producing people in this fashion. *Brave New World* may not be science fiction much longer.

We're not through with this monster that biological scientists have the potential to make. Remember Dr. Petrucci, the Italian scientist who was driven out of Italy by the Vatican? Using his knowledge, Russian researchers were said to have about two hundred fifty human fetuses growing in artificial wombs at the Institute of Experimental Biology in Moscow. The Russians clamped a tight news lid on their experiments and did not permit pictures or interviews by Western journalists. They may have produced some malformed babies, which they didn't want the world to know about.

Artificial wombs that carry babies to term will probably be available early in the new millennium, as science moves closer from both ends of the gestation process. More fully developed embryos are being grown in the lab, and special environments for "preemies" with hyaline membrane disease are already in the hospital nursery. When the technology for growing embryos and saving preemies meets, we'll have artificial wombs. Then a woman won't have to hire a host mother. She can donate her egg and let a machine do all the work. Childbirth for the rich will be painless. They can leave sperm and eggs at a lab, go about their regular business, and drop by the lab or hospital in nine months to collect their baby.

Dr. Leon Kass calls the development of the test-tube baby "dehumanization" of the birth process. "Quality control of the product," he says, is not worth "the depersonalization of the process." He asks, "Is there not wisdom in the mystery of nature that joins the pleasure of sex, the communication of love, and the desire for children in the very activity by which we continue the chain of human existence?"[8]

7

Artificial Insemination

The science of producing test-tube babies is now two decades old, and there are many of these babies living at this writing. Some, in fact, have two "fathers." The biological dad, whose sperm fertilized the mother's egg, is almost always off the scene when the baby is born. He was only needed for fertilization of the mother. The other father's name is on the birth certificate, just as if the child is his hereditary son or daughter. (Adoptive parenthood is quite different. There, neither of the parents who claim the child as their own holds hereditary kinship with this new member of their family.)

In the Artificial Insemination by Donor (AID) procedure, the physician simply inseminates another man's fertile sperm directly into the woman's uterus, and nature takes over from there. Because of the secrecy involved, no exact tabulation is available on the number of AID offspring alive and well in the United States. Estimates within the U.S. medical profession run from two hundred thousand up to five hundred thousand with perhaps twenty thousand new AID babies being born each year. While Artificial Insemination Husband (AIH) is often freely talked about among family and friends, AID is not. Husbands do not want their sterility known.

Of course, this is not cloning, where the baby has only single-sex heredity. An AID child is always a genetic mix from a man and a woman. A clone may grow up in a home where there are two heredities. But the clone will be related to only one, the person from whom he or she has been cloned.

AID has become a routine procedure for many doctors. The physician schedules a woman's appointment for the time in her menstrual cycle when she is most likely to conceive. When his patient arrives, he is ready with a specimen of donor sperm that may have been stored in a sperm bank. The doctor merely inserts some of the sperm into the woman's cervix with a small syringe, then tucks inside her vagina a tiny plastic cup containing the remaining sperm. The doctor instructs her not to swim or douche during the next six to eight hours; after that, she should remove the cup. If the first attempt doesn't succeed, she returns about a month later for a second attempt. Most healthy women conceive after one or two inseminations with fresh sperm from a donor; frozen sperm from a bank usually requires three or four treatments.

AIH may be employed when the husband is not sterile but is unable to impregnate his wife through ordinary intercourse. He may be paralyzed, for example, or his sperm count may be too low. In cases like this, the physician will take several collections of sperm and inseminate with the concentrate. Or there may be some chemical incompatibility between the husband's sperm and the acid secretions in his wife's vagina. The doctor will then take a sperm specimen and bypass the trouble spot by inseminating the sperm directly into the oviducts. Another possible use of AIH is when the husband wishes to have his sperm frozen for later insemination. He may be going off to war or undertaking a dangerous job assignment. He may have to undergo surgery that will render him sterile. Or he may wish to have a vasectomy. By storing his sperm, he can father a child at a future date of his and his wife's choosing.

The propriety of AIH is questioned by some Roman Catholic authorities who judge the rightness of the procedure by the method in which the sperm is obtained. They say the act is licit if the sperm is obtained during intercourse, but it is wrong if secured by masturbation. Protestant theologians generally do not object to AIH in any form. Medical specialists tend universally to recommend AIH as a worthy means of conceiving children. Dr. S. J. Behrmann, former director of the Center for Research in Reproductive Biology at the University of Michigan, recalls helping a husband who was paralyzed and could not have normal sexual relations with his wife. "A year or so later, I saw this man in his wheelchair holding his child. He was crying for joy."[1]

AID, as I have previously noted, is much more controversial. Some doctors who do AIH will not take an AID couple. Some doctors who will accept an AIH married couple will not administer donor sperm to a single woman who just wants to have her own baby. AID raises many legal, moral, and medical questions. Let's look first at the history of AID.

The History of AID

The practice of artificial insemination began as an aspect of animal husbandry. One of the first accounts in ancient literature tells of an Arab sheik, living around

A.D. 1000, who inseminated his enemy's sleek mares with sperm from sickly stallions. A few years later the inseminator won an easy military victory. Another historical example occurred around 1780, when an Italian physiologist, L. Spallanzani, inseminated a dog to produce a litter of three pups.

England claims the first human birth by artificial insemination. About 1800 Dr. John Hunter successfully inseminated the wife of a London linen merchant. The first American AID success was reported in 1866, when Dr. J. Marion Sims claimed that six of his women patients became mothers through donor sperm. He didn't name the women. AID was a well-kept secret within the medical fraternity before World War II. Today, when almost anything can be printed or shown on television, it is difficult for us to imagine a time when words such as *womb, ovary, sperm, oviduct,* and sometimes even *pregnancy* were taboo in public newspapers. Bearing a child out of wedlock was a scandal then too. For fear of their reputation and also of prosecution, doctors didn't keep records of donors and recipients. The legal husband's name always appeared on the birth certificate, even though he was not the biological father. We will never know how many hereditary family lines have been broken by AID.

Cattlemen did keep records. In the 1930s the first AID calves were born in Wisconsin. Half of all American dairy cows are now sired by sperm taken from prize bulls. Dairy farmers employing AID have increased milk yield up to 65 percent by selective breeding. Crossbreeding of animals is now common around the world. Peruvians, for example, have crossed llama males, which once provided clothing for Inca royalty, with alpaca females. The result is an animal with a fine grade of fleece.

Animal and human AID would not be possible on a large scale today without the development of deep freeze sperm banks. Again, the English take the honors in pioneering. In 1959 an English scientist, Dr. C. Polge, found that he could protect frozen sperm by adding a quantity of glycerol. Semen, containing sperm, is mixed in a twelve-to-one ratio with glycerol and kept in a liquid nitrogen freezer at $-196°C$. The sperm is thawed by merely exposing it to room temperature. Large livestock breeders now routinely use sperm banks for improving their herds. Sperm from one bull may be used to produce a thousand calves in one year.

Human AID practitioners in the United States serve their patients from a number of sperm banks as well. I have heard of only one reported mishap. A freezer failed, resulting in the loss of hundreds of specimens. Thirty-five clients, including some single men who had had vasectomies and cancer patients who had stored sperm while undergoing radiation treatments, sued the bank's owners. There have no doubt been other "accidents."

Frozen sperm can be kept potent for up to ten years, although samples more than two years old are seldom used. Some doctors disdain the sperm banks and will use only fresh semen, which they claim gets better results. Some thirty years ago, Drs. Keith Smith and Emil Steinberger of Houston recorded a conception rate in their patients of 61 percent with frozen sperm, compared to 73 percent for fresh semen.

The current payment for a sperm specimen varies. I have never seen a list of fees. Medical students, interns, and residents are preferred donors. They are young, easily available, usually need the money, and are considered healthier than men in the general population.

A reputable AID physician will counsel a couple seeking AID. The physician wants to know how badly the couple wants a child and if they are mature enough to handle psychological problems that may come up. The doctor's fee is covered by some insurance companies as a surgical procedure, but to my knowledge, AID is still not available to charity patients except for voluntary experimental purposes. This has brought the complaint that AID is only for those who can afford it. Whether they use an immediate donor or a sperm bank, AID doctors try to select sperm from men who match the husband's physical appearance and blood type. The idea is to make everyone, even the grandparents, believe that the baby is really the husband's child.

The Myth of Donor Screening

Many couples seek AID because of fear that their own union could produce a child with genetic disease. Until recently it was presumed that all doctors ran careful heredity checks on sperm donors. A University of Wisconsin survey of 379 AID physicians—the most comprehensive survey of its kind ever conducted—reveals that many do not. Only 12 percent of the physicians replying to the questionnaire said that they did chromosome tests on donors to prevent birth of a Down's syndrome child. Only 30 percent admitted checking for traits that might result in other serious genetic defects. Ninety-four percent said they would not use a carrier of Tay-Sachs disease as a donor, yet fewer than 1 percent tested for this affliction.

More than two-thirds of the doctors replying confessed that they kept *no* files on donors. Some had used the same donor for several pregnancies; one for six was not unusual, and one doctor reported the "fathering" of fifty children by a single donor. This raises the specter of half brothers and sisters innocently marrying one another, thus increasing their chances of bearing diseased children. In 1974 an engaged couple who grew up in the same town canceled marriage plans after learning from their family doctor that they were actually half brother and half sister.

Who Else Is Using AID?

Forty-seven physicians in the Wisconsin study said they had inseminated single women, including lesbians, on their request. The doctors who administer AID are caught in a bind. Some—the majority, I would hope—do not want to help unmarried women, particularly lesbians, have children. The tide of public opinion in this country is in favor of freedom of choice. Therefore an AID doctor may not be able to refuse the request of any woman for insemination, as long as she can pay for it. If a woman can get an abortion on demand, I don't see how AID can be withheld from lesbians.

I would like to pause to mention a technique that, if perfected, would appeal to lesbians more than AID. Dr. Pierre Soupart of Vanderbilt University Medical School has been developing a method of starting a new life with two eggs rather than an egg and a sperm. The offspring would have two biological mothers and would be a female. "It's the answer to a number of women's dreams," says Lucia Valeska, coexecutive director of the National Gay Task Force. Each woman would donate an egg that would then be treated in such a way as to cause the eggs to fuse into one cell. That cell would be comparable to the zygote that results from the fusion of a sperm and an egg. After a few cell divisions in the laboratory, the growing embryo would be implanted in the uterus of one of the mothers. Dr. Soupart has carried out this procedure at least once with mice and has plans for further improvement of the technique. Valeska says, "It's very understandable that women who want to raise children together should want the children to be part of their biological makeup."[2] In my judgment, egg fusion is a radical departure from God's plan for human reproduction. Because it is not yet available, we can only speculate on the effects of its use. But AID is with us now, and the questions it raises are of immediate concern.

Effects of AID on Families

In addition to the use of AID by lesbians, controversy centers on possible psychological and social problems in an AID family. Does having an AID child help or hinder a marriage? Dr. Behrmann, who formerly worked at the University of Michigan, knows of only one divorce among six hundred recipients. Dr. Sheldon Payne of the Shelton Clinic in Los Angeles says he has studied AID couples over a span of thirty years and found their divorce rate to be one-fifth of the California average. He admits this may be because of the way couples are screened before being accepted for AID at his clinic. Having a child to stabilize a marriage, he says, is an immediate disqualifier. On any practice as controversial as AID, you can usually find statistics and stories to support both sides, depending on who is doing the talking. I don't doubt that Drs. Behrmann and Payne are telling the truth about their patients, but what about the patients of doctors who are not so conscientious?

I won't pass any judgments here, except to note that one medical researcher has discovered some bad results of AID. He is Dr. Laurence E. Karp, codirector of the Prenatal Diagnosis Center of the Department of Obstetrics and Gynecology, Division of Medical Sciences of the University of California at Los Angeles (UCLA), located at Harbor General Hospital in Torrance, California. The following examples are from his book *Genetic Engineering: Threat or Promise*.[3]

- In two cases where AID was urged by doctors on couples, the husband of one and the wife of the other suffered serious mental depression afterward.
- A British physician gave up AID because he perceived that the practice caused husbands to feel inferior and the wives to have undesirable mental attitudes toward the anonymous donors.
- The depression experienced by five husbands from not being able to father children was worsened by their wives having children by AID. The wives reportedly had sexual fantasies about the unknown donors and the doctors and rejected the babies prior to birth. After birth, all of the children in these five families suffered severe emotional damage.

These may represent only a small percentage of the thousands of couples who have opted for AID, but they should make any couple or counselor think a bit before resorting to the procedure.

Legal Entanglements

The legal problems are also troublesome. To my knowledge, only four states—California, Georgia, Oklahoma, and Kansas—have laws legitimizing AID offspring, and then only if the husband has consented to the insemination. The Oklahoma law stipulates that a couple desiring AID must go before a county judge and sign consent. This is filed with the court and is similar to adoption. The AID child, when born, holds the same rights as if he were the couple's natural child. One other state, Arkansas, has a law protecting inheritance rights of AID offspring.

Results of the few court cases related to AID have been contradictory. The first North American legal battle occurred in 1921 and resulted in an Ontario, Canada, court deciding that AID constituted adultery. The first United States case, occurring in Illinois in 1948, resulted in a different opinion. The judge said AID could not be adultery; then he granted the husband a divorce because the wife had engaged in regular adulterous liaisons. The year before, a New York court also held that AID was not a form of adultery and ruled the child legitimate, provided the husband had consented to the procedure. The judge said the child had been "potentially adopted or semiadopted" by the father and, therefore, the father had the same rights that a foster parent would have. The mother later moved to Oklahoma where she was granted a divorce and custody of her child.

In a 1954 divorce case heard by the Superior Court of Cook County, Illinois, the judge noted that an AID child had been born in lawful wedlock. The next year this court reversed itself and ruled that AID "with or without the consent of the husband, is contrary to public policy and good morals, and constitutes adultery on the part of the mother. A child so conceived is not a child born in wedlock and therefore is illegitimate."[4] Few AID-related cases that I know about have come to court since 1954. None clear up the legal thicket in which AID participants are caught in most states, and the Supreme Court has never ruled on the subject.

Thousands of children have been fathered through AID, but the parentage of only a few is known. The usual custom is still for the husband to be declared the biological father. Sometimes the obstetrician knows the truth yet falsifies the birth certificate to avoid trouble. Sometimes the doctor doesn't know and innocently misrepresents the parentage. In many instances, a couple will go to a fertility specialist in another city. The woman receives donor insemination, returns home, and shows up at her doctor's office pregnant. I can understand why the husband is declared the biological father in most situations and why doctors keep such poor records. The potential for trouble is great. A husband may decide to sue a donor for alienation of affection. An AID child may demand participation in his or her biological father's estate. Should the husband and wife die, relatives may demand that the AID child be disinherited. Any lawyer could suggest a dozen more possibilities for court action.

A host mother, who carries to term another's test-tube baby, will present the potential for similar problems. More court cases may result from the use of host mothers for the simple reason that it is easier to conceal a sperm donation than a pregnancy. AID and host mothers also pose problems in the moral arena. Does the bringing of a sperm donor or a host mother into a marriage constitute adultery? One may argue that the use of a host mother does not constitute adultery, since she herself would not conceive the child.

Moral Dilemmas

Let's look further at the moral question: Is AID adultery? There has been a lot of discussion on this in church circles. Catholic authorities continue to hold that AID is adultery. Fr. Francis Filas, professor of Theology at Loyola University in Chicago, bases the following view on "principles of official teaching."

> Adultery is the violation of the marriage bond which is oriented to new life and in which husband and wife have a right to each other's life-giving powers. The husband's right to his wife's procreative powers is a gift from God which he cannot give away. He can no more tell his wife to receive semen from a donor than he can tell her to have intercourse with another man.[5]

Protestant denominations appear to be divided on whether AID constitutes adultery or is no more immoral than any other medical procedure. The United Presbyterian Church endorsed a report in 1962 that accepted AID with the qualification that doctors should be sure of the "intelligence and emotional stability" of couples before advising this "radical social procedure." But the Presbyterians refused to call AID adultery. This, they said, would give adultery "a meaning it does not have in the New Testament." In 1970 the Lutheran Church of America declined to interpret AID as an act of adultery. "Christian ethics," the Lutheran theologians said,

> cannot take a categorical position disapproving of artificial insemination by donor. Nevertheless there are psychological, social, and legal reasons which might lead Christians to refuse artificial insemination by donor for themselves. Finally, of course, the decisions rest with the persons who are involved.

Other statements by Mainline Protestant denominations run about the same. I know of no conservative evangelical group that has spoken on the matter. When questioned on the subject, Billy Graham responded, "The procreative process was devised by an infinite, wise God, and is . . . sacred. The Scriptures mention no other method of reproduction of the human species. . . . When the Scriptures are silent, there is a show of doubt."

In England the Anglican Archbishop of Canterbury appointed a special commission of theologians to study AID. The result was this statement:

> The strongest plea put forward in defense of AID is the plight of the married women who long for children but whose husbands are sterile. . . . We need not affirm once more the profound compassion which this frustration must evoke; but our concern at this point is to answer the question whether such a desire . . . [is] in strict truth inordinate: one which exceeds the proper bounds of de-

sire. . . . We are all, without distinction, required to re-
strain our desires, however imperious. On what rational
ground is it urged that while sexual desires ought not to
be indulged at all, parental desires may be?[6]

Helmut Thielicke, a European Protestant theologian,
is likewise opposed to AID. He agrees with Catholic
moralists that marriage conveys exclusive bodily rights
to sexual and reproductive organs and that AID would
be a violation of these rights. He also claims that AID
threatens marriage because it fulfills motherhood while
confirming failure of the father. Fidelity in marriage, he
insists, is first a personal bond between husband and
wife and not primarily a legal contract. Parenthood is
"a moral relationship with children, not a material or
merely physical relationship."[7]

As we might expect, situational ethicist Joseph
Fletcher sees no moral problem with AID. For one who
denies the authority of Scripture, he takes the curious
stance that AID is permissible because polygamy, con-
cubinage, and levirate marriage occurred among God's
Old Testament people. He notes that Leah and Rachel
sent their husband Jacob to conceive children by their
handmaids (Gen. 30:1–13); barren Sarah dispatched her
husband, Abraham, to her fertile maid, Hagar (Gen.
16:1–2); and Onan displeased the Lord by refusing to
impregnate Tamar, the wife of his dead and childless
brother, Er (Gen. 38:8–10).

Other reasons besides adultery make AID questionable. One is the deceit involved in naming the husband
as the father on the birth certificate. I'm old-fashioned
enough to believe that a lie is a lie. Furthermore, it's
wrong from my viewpoint to deceive an obstetrician,
who does not know what has taken place, into thinking
that a woman is pregnant by her husband so that he or
she will declare the husband the father.

Other Options

To a Christian couple contemplating AID, I would say, "Have you considered that it may be God's will for you not to have children of your own? Might he not have an adoptive child or a foster child for you to love or some other outlet for the care and time that you would give a natural child?"

I know that it isn't easy to find a child for adoption. Millions of babies who might have been adopted have instead been aborted. And then most unwed mothers who choose not to abort are keeping their babies because the public stigma is not as great as it once was. Still, I think a couple ought to look into adopting an American or foreign child. It may be expensive, but look at what it costs to have a child in a hospital. To find an adoptive baby, you need to keep your eyes and ears open. Ask friends to be on the lookout for a prospective baby. In 1994 the mother of a young husband in Tennessee just happened to be working for two obstetricians. The husband asked his mother to be on the lookout for a prospective baby. The mother found two pregnant women caught in different marital situations. Arrangements were made for their babies to be adopted at birth. The Tennessee couple now has two beautiful babies, and their go-between has two grandchildren.

If adoption isn't possible, there is the possibility of caring for a foster child or of sponsoring a child overseas through a missionary agency. These can be very rewarding endeavors.

The late Dr. H. J. Muller, a Nobel prize winner in physiology and medicine, called on prospective parents to forego "egotistical" desires to reproduce their own heredities and, instead, help advance the human race by constructing children from the "best" available egg and sperm.[8] This is the same old song of humanistic evolution. With the development of IVF and the prospect of cloning added to AID, we're now hearing more along this line. The Muller plan is to store the best sperm and eggs in banks until the political climate permits establishment of experimental farms. A California sperm bank containing the seed of several Nobel prize winners was recently named for Muller. His widow has protested the use of her husband's name.

I can deal only briefly with the use of donor eggs in reproduction. For donor eggs to be implanted in a woman's ovaries, she must first have injections of hormones that cause her ovaries to swell. She must also submit to blood tests and ultrasound examinations for monitoring the development of the eggs.

The first donor eggs were used in 1984. At that time, according to Dr. Mark Sauer, a fertility specialist at Columbia Presbyterian Medical Center in New York, donor eggs could be bought from donors for $250 each. Today donor eggs are harder to come by. Due to the scarcity, St. Barnabas Medical Center in Livingston, New Jersey, is offering to pay $5,000 an egg.

Frightening? You bet. But nontraditional methods of reproduction through cloning, IVF, and AID are only part of the picture. In the next chapter we'll start delving into the larger world of genetics and look at procedures that could be used to change human heredity itself.

8

The Heredity Trap

A child bounces happily on her father's lap. The mother looks on fondly at her two look-alikes. The freckles on the bridge of their noses are the same. They have the same jutting chin, even the same quick laugh. Credit heredity.

Some cloning supporters see examples like this as a larger picture in which the best heredity is multiplied through multiple clones. However, this is not the whole picture. Heredity has a dark, painful side to it too.

A two-year-old sightless child screams in pain. His parents try to comfort him while the nurse is preparing a narcotic. Their faces are strained, reflecting months of mental anguish. Their child is a victim of hereditary Tay-Sachs disease, which afflicts an average of one in thirty-six hundred Jewish boys. Merciful death usually comes before age three.

Huntington's chorea is also inherited. Because it appears later in life, many victims already have children by the time the disease appears. The children must then live with the agonizing uncertainty of whether they, too, will have the disease. About twenty-five thousand Americans suffer from this affliction. Victims lose control of facial and body movements, one muscle at a time. They frequently lose their balance and fall. Their arms begin flailing at the most embarrassing moments. Severe emotional problems, depression, and violent behavior may result. Many victims commit suicide.

Muscular dystrophy, marked by progressive weakening and degeneration of muscle fibers, affects about two hundred thousand children and youth in the United States. It is actually a class of diseases, the most common being Duchenne dystrophy, which is thought to be inherited by boys from their mothers, who themselves do not manifest the disease. The victim appears normal at birth, but before six he is waddling, walking on his toes, and falling frequently. By twelve he will probably be in a wheelchair. By twenty he will likely be dead from a respiratory infection his breathing muscles were too weak to combat. Millions of dollars are spent each year on physical therapy, drugs, and research related to Duchenne dystrophy and thirty-seven other neuromuscular diseases. No miracle cure has yet been found. At best, only certain symptoms can be controlled and victims helped to live as normal a life as possible. Much is still to be learned about the origins of the various dystrophies. The consensus among researchers is that heredity is probably at fault.

Take another disease that comes through the victim's genes. Here's a young man of twenty-eight, father of two, just starting his career. He loves sports and is the picture of health. One day he collapses while playing tennis. His partner speeds him to a hospital, where the doctor pronounces him dead of coronary thrombosis. A coronary artery had become hard and narrow because of arteriosclerosis. A blood clot had blocked the artery, cutting off the blood supply to part of the heart. Arteriosclerosis—caused by the buildup of fatty deposits and calcium—is not unusual in older men. But why is it present to such a large degree in one so young? Doctors speculate that the young father suffered from hypercholesterolemia, a genetic disease that involves a deficiency in the chemical agents that normally hold down cholesterol. Left unregulated, the cholesterol caused fatty deposits to build up rapidly in the arteries.

Each of us begins life as a combination of the heredity of our parents, which they in turn inherited from their parents. Most of us are born relatively healthy. I say relatively because no one escapes a hereditary handicap in some form. It may be as harmless as flat feet or as devastating as SLE (systemic lupus erythematosus, a connective tissue disorder of internal parts of the body). One may have a predisposition to a more familiar disease, such as arteriosclerosis or emphysema, or perhaps a combination of factors that will never be diagnosed by medical science. I'll say it again: No one is born in perfect genetic health. Each of us inherits some bad heredity due to mutations that have accumulated during the history of the human species. As of 1997, doctors have tests that can identify 450 genetic diseases. New hereditary afflictions are discovered almost every week. The newest discoveries include Fragile X syndrome, a common cause of mental retardation, and neurofibromatosis, a rare disease causing disfiguring tumors.

When you hear the term "inherited disease," you may assume the problem is too complicated to understand. However, the basic principles of genetics are not all that hard to grasp. The exceptions and the combinations of factors involved may be baffling, but this shouldn't keep you from understanding how the laws of heredity ordinarily work. Is it really essential to know about chromosomes, genes, and DNA when genetic disease hasn't touched you or one of your loved ones? It is if you want to understand what some of today's scientists have in mind for reshaping the human race. If you're interested in the heredity of your own posterity and why many scientists want to change it, you'd better care. You may decide that the new genetic engineers are instruments of mercy. Or you may conclude that their research testing on humans, even with the best of intentions, is diabolical and should be banned altogether. Closer to home, you need to know all you can about genetics, because one or more hereditary problems may pop up in your family, perhaps through an in-law who has entered your circle through the door of marriage.

A Crash Course in Genetic Inheritance

Class is back in session. This time, we'll start with a history lesson.

Some History

The father of modern genetics was an Austrian botanist and Roman Catholic monk named Gregor Mendel who lived in the nineteenth century. Before his discoveries were made known, people could only speculate about how family traits were passed on. Aristotle, for example, believed male sperm contained all the plans for an offspring. He was at least half right. Mendel proved, among other things, that the characteristics of offspring are shaped by both parents and follow a predictable pattern, generation after generation.

Mendel discovered that crossing yellow-seeded peas with green-seeded ones always produced a totally yellow first generation. But the green color always returned in the second generation, in one of every four offspring. This led him to two conclusions: (1) Certain characteristics are dominant over others when different parents are crossed, yellow over green, for example, with the characteristics of only one of the parents showing up in the offspring. (2) Other characteristics are recessive, or secondary, but they are not wiped out in the crossbreeding. They reappear in succeeding generations. Mendel confirmed this in one experiment after another. He showed that the reappearance of characteristics, or hereditary traits, could be predicted mathematically, generation after generation.

Mendel's findings were printed in a scientific journal in 1866, but they were ignored by scientists—perhaps because the monk's discoveries contradicted the theories of another scientist named Charles Darwin. Darwin, whose book on evolutionary theory was published in 1859, believed that acquired characteristics could be inherited and that change was a normal process. By Darwin's reasoning, the giraffe's long neck was "acquired" from stretching to reach leaves on high trees. This evolved, in time, into a hereditary pattern.

While Mendel's discoveries gathered dust, western scientists were coming to believe that heredity was transmitted through the sperm and egg that fuse to produce the first cell of a new life. By conducting experiments with fruit flies, Thomas Morgan unknowingly confirmed many of Mendel's conclusions, while demonstrating that exceptions sometimes did occur in the mathematics of transmission. Mendel's report was finally "discovered" about 1900. The precepts the monk had so carefully set down demonstrated what many western scientists now suspected. He became a scientific hero in the United States.

It was a different story in the Soviet Union. A botanist named Lysenko convinced Stalin and his Communist Party bosses that Mendel's laws were all a part of an imperialist plot. Acquired traits, said Lysenko and his associates, were transmitted to offspring. If you cut the tails off dogs for enough generations, Lysenko predicted authoritatively, a dog would eventually be born without a tail. The ideas of the "pea-picker" (Mendel) and the "fly-breeder" (Morgan) were officially condemned by the Central Committee of the USSR's Communist Party. Because of such dogmatism, the USSR would remain far behind the West in knowledge of genetics. This vastly affected their method of agriculture and partially contributed to their continued poor grain harvests. Since the breakup of the Soviet Union, Russia and other now-independent states have been trying to catch up with the United States and western Europe in plant and animal breeding.

Modern Advances in Genetic Links to Disease and Defects

Our knowledge of human genetics keeps advancing rapidly. As I noted a few paragraphs back, scientists now have tests for about 450 genetic diseases. Some of these represent varieties of cancer. Researchers at the Creighton School of Medicine in Omaha, Nebraska, discovered that between 10 and 20 percent of all varieties of cancer are transmitted from generation to generation. Breast cancer heads the list. Genetic diseases that are now known can roughly be placed in three categories.

First are those diseases caused by abnormal numbers or shapes of chromosomes. Chromosomes, you'll remember, are found in the cell nucleus, twenty-three pairs to a set, with each parent contributing one in each pair. An example of this type is the familiar Down's syndrome. The cause of this affliction was discovered in 1959 by Jerome Lejeune, a French geneticist who later crusaded against abortion. In fact, two months before Lejeune died in 1994, Pope John Paul II named the geneticist to lead the Pontifical Academy for Life, dedicated to promoting Catholic teaching against abortion. Down's syndrome usually results from having an extra twenty-first chromosome, three instead of the usual two. This is more likely to occur with the child of an older mother. We don't quite know why, although recent research relates it to the decrease of female hormones. Other research has shown that the risk is also increased for older fathers.

Another defect from chromosomal abnormality is Turner's syndrome, which prevents a child from developing secondary sex characteristics at the age of maturity. This disease is marked by the presence of a single X chromosome, with no X or Y partner to produce a normal female or male. Another condition that affects sexual development is Klinefelter's syndrome, in which the cells have two X chromosomes plus a Y chromosome.

The second group of genetic diseases are those caused by mutations, chemical mistakes in the construction of individual genes. (Genes, as we've already said, are contained in chromosomes and are matched in pairs accordingly.) Many diseases fall into this category: hemophilia and Duchenne dystrophy are just two. Another is PKU (phenylketonuria), a metabolic disorder that causes mental retardation unless treatment begins shortly after birth. Another is galactosemia, which prevents a baby from metabolizing the milk sugar, galactose. If given milk, the child may die within two weeks. A simple change of diet will prevent tragedy.

The third class of inherited diseases are those that are influenced by a number of genes and may be set off or pushed forward by environmental factors. Hypercholesterolemia, for example, is aggravated by a diet rich in cholesterol. A low-cholesterol diet may prevent or at least postpone a coronary. Also, researchers at a Veteran's Administration hospital found that 50 percent of their emphysema patients had an inborn shortage of the biochemical alpha–1-antitrypsin, a chemical that ordinarily fights pollution and other foreign substances in the lungs. People with this deficiency who protect their lungs against cigarette smoke and other dangerous pollutants may never suffer emphysema. Thirteen times as many smokers as nonsmokers develop emphysema. Their lungs become less elastic, until finally they are literally unable to breathe.

We're discovering more and more links between some genetic diseases and environment. Some scientists think that homosexuality results from a combination of genetic irregularities and early life experiences. Other researchers think that "gayness" may be mostly genetic. Still others believe homosexuality is triggered completely by experience. It is only clear that the cause is not yet clear!

USA Today, the best-known newspaper in America, ran a front-page article titled "How Much Do You Want to Know?" (about the genes of unborn children). The *USA Today* writer says flatly, ". . . today there is no prenatal test for gayness."[1] Ignoring this fact, the *USA Today* writer observes, is a cover of a recent issue of *The Advocate,* a magazine for gays and lesbians. The cover shows a picture of a fetus under the headline: "Endangered Species: This Child Has the Gay Gene. Will He Be Aborted Because of It?" The *USA Today* article notes further falsehood in a gay-oriented 1993 Broadway play, "The Twilight of the Golds." A woman learns through

114

prenatal testing that the baby she is carrying will likely be gay. The play ends with the woman aborting the unborn child.[2] These are two examples of pure propaganda. I suppose if gay writers and their sympathizers say it enough, many readers will come to believe it and not bother checking with unbiased authorities. Although no homosexual mutation has yet been identified, it is true that we're discovering more and more links between certain genetic diseases and environment. Applying the knowledge for this field could increase life expectancy several more years.

The Downside of Genetic Advances

Some scientists, however, are not pleased that modern medicine is learning how to control genetic disease. Why? There was a time when people with genetic diseases didn't live long enough to bear children. Consequently, they didn't pass on their problem heredity. Today, however, they're living longer and raising families. Their bad heredity is multiplying and spreading, resulting in an increase of genetic defects in the population.

This problem has also been aggravated as medical science has wiped out or effectively curtailed one killer disease after another. More soldiers, for example, died from typhoid fever during the Civil War than were killed in battle. When did you last hear of someone dying from typhoid? The flu epidemic of 1918–19 took more lives than World War I. In one recent year only about seven thousand influenza deaths were recorded, and most of these were already fragile elderly people. Typhoid and influenza are not genetic diseases, but these and other nongenetic maladies tend to take a greater toll among persons with hereditary deficiencies.

To illustrate further: Once people born with poor or failing eyesight had a hard time surviving. Some got picked off by wild animals. Some may have been run over by chariots. Now we have eyeglasses, contact lenses, and remedial surgery. Most people with bad eyes live and enjoy life as long as those with good eyes do—and they keep supplying the bad genes to the population. This is good for optometry but bad for genetics.

A cold-blooded scientist will argue for a population control that calls for killing off the genetic weaklings or sterilizing them so they can't bear children. Some would say that cloning is the perfect remedy for this problem. Clone those with the best heredity. Abort all fetuses who do not pass a genetic test. This is the only way to build a healthy race. So they say. A totalitarian government could, I suppose, do this. The despots in power would appoint a team of scientists, a panel of Dr. Deaths, to identify the undesirables. Could this happen? Of course. Can you imagine what Hitler and his team of doctors would have done with the genetic knowledge now available?

In a sense, though not with genetic testing at the root, killing undesirable offspring has already happened in some societies. One example of such a society is the Aucas, the Indian group that in 1956 killed five American missionaries trying to contact them. The Aucas were then probably the most isolated tribe in South America. No outsider had ever lived among them.

Much has happened since the five missionaries were killed. Elizabeth Elliot, the wife of one of the martyrs, and Rachel Saint, the sister of another, entered Auca country with a young Auca woman who had come to the outside. The very men who had killed the missionaries were converted, and most of the tribe are now Christians. Before Christianity came, though, it was the custom of this small tribe to kill weak babies, usually by choking them to death or burying them alive. The missionaries taught them to spare these little ones. The "weak" children who have been saved are now old enough to have grandchildren. There has been no noticeable genetic change in the tribe as yet, making it possible for medical researchers to determine the well-being of a group that had habitually killed off weaklings.

A team of doctors from Duke University Medical School visited the Aucas and discovered no evidence of hypertension, heart disease, diabetes, or cancer. The highest blood pressure they could find was a systolic reading of 109 millimeters in an older man who was upset about something when his pressure was taken. A reading of 120 is considered normal in our population for a twenty-seven-year-old male. The Aucas will probably not remain this healthy for long, though. The weak ones will have children, thus spreading bad heredity in the group. The tribe has already been exposed to civilized diseases. In 1969 a polio epidemic swept through the tribe, killing sixteen. A Quichua Indian from the outside was blamed for bringing the disease. The Aucas have since been vaccinated by missionaries, who are extremely careful about exposing any of them to infectious maladies.

Looking at the Auca experience from a scientific point of view, does it make sense for the missionaries to be there? Many non-Christians will say no. But there are a couple of other factors I haven't mentioned. First, the Aucas and other "primitive" tribes are going to be exposed to civilization, whether the missionaries go to them or not. It's just a question of who is going to help them face the modern world. Second, it has been determined that before the missionaries came, the Aucas had a high death rate. More than 50 percent of this mortality resulted from spearings in revenge raids. Since missionaries came, these raids have stopped, and the population has more than doubled. So, at least with the Aucas, we must consider more factors than just genetics.

Genetic Armageddon?

Pessimistic scientists, however, are looking at all the "tribes" of the world from a wider genetic perspective. They say that the way humanity is going (i.e., saving lives and allowing genetic defectives to live and reproduce), we're headed toward a genetic Armageddon. They predict this from simple laws of probability. With more people alive carrying bad genes, the likelihood of disastrous combinations in marriage grows, resulting in more diseased children.

Let's simplify this a bit: A hundred years ago, ten people in a thousand may have carried bad genes for a certain genetic disease that would not appear until two carriers married. The chances of this happening were almost infinitesimal. Now the ratio of carriers to population may be fifty in a thousand, increasing the probabilities greatly. But I see a counterpoint. Centuries ago, world population was much smaller than it is today. Lack of good roads and efficient means of transportation kept people from traveling far from home. Fewer marriage prospects were available. Many cousin marriages occurred, and relationships lacked the genetic variety needed to maintain good heredity.

My colleague in writing, Dr. James C. Hefley, was raised in an out-of-the-way valley down in the Ozarks. His parents were third cousins. One of his first cousins married one of his mother's sisters. He has found other marriage linkages among his kin. When Jim's parents were young, people rode a horse or wagon or walked to get where they wanted to go—which wasn't far, Jim says. By the time Jim finished high school, the roads had been improved, and his father had a Model-A pickup. Jim went to college two hundred miles away. After getting his degree, he moved to New Orleans, where he met Marti Smedley, a native of northern Michigan. Among more than sixty first cousins, Jim was the second to marry someone from outside the valley in which he was raised. As a result, his and Marti's descendants are more likely to have better genes than if they had been cousins in a large clan.

Good Genes and Bad Genes

Let's look at some basic facts about how bad genes can interact in a marriage to produce a child with genetic disease. Genes, as we've said, come in pairs. In each pair one gene is dominant and the other recessive. Let N represent a dominant healthy gene, n an abnormal recessive gene. A pair of two normal genes is written NN. A person with such a pair is called a homozygote (*homo* means "same"). A normal and an abnormal gene in a pair would be written Nn. A bearer of this combination is called a heterozygote (*hetero* means "different").

The marriage of two normal homozygotes (NN plus NN) almost always means that all their children will be NN. Only in very rare instances will outside influences (such as radiation) damage the character of a gene, changing a good N into a bad n. However, the marriage of two heterozygotes can mean trouble. For example, in Mendel's crossing of garden peas, when dominant yellow (Y) and recessive green (y) are paired in both parent pea plants (Yy), there is one chance in four of producing a green (yy) offspring. Peas and people act alike in this regard. If a man with Nn marries an Nn woman, each child will have one chance in four of receiving two harmful ns. Without a good N to control the bad n, a defect will result at this spot in the chain of inheritance.

Let's apply this to cystic fibrosis, a disease that causes malnutrition and an excess of mucus secretions through-

119

out the body. Its victims often die from lung infections. One in every twenty-five people in our population carries one of the genes that causes cystic fibrosis. This person is Nn, the little n standing for the bad gene. If the carrier marries a normal person, the formula for each of their children will be NN or Nn, and none of their children will have the disease. The dominant N will checkmate the recessive n. But suppose a carrier marries a carrier (Nn + Nn). In this case, one out of four of their children (on average) will be born with cystic fibrosis.

Or consider hemophilia, one of the so-called sex-linked diseases. It's called that because it is the result of a gene carried on the X chromosome. The cells of a female each have two X chromosomes, while those of a male have one X and one Y chromosome. A female with a bad sex-linked gene will have one X that is n and a healthy X with N. The healthy N would counterbalance the bad n. She will only be a carrier. But the male has only one X chromosome, so if he inherits the bad n, there will be nothing to counteract it, and he will have the disease. His Y chromosome does not contain the same gene.

All of this may seem like an alphabet game, but it is agonizingly serious to families who have children suffering from genetic diseases. It happens in tens of thousands of families every year—rich and poor, mighty and lowly, just and unjust. Genetic diseases play no favorites and operate according to the laws of heredity, which are rarely fickle. It is getting worse because bad genes are increasing faster than the population, since carriers live longer and pass on their potentially lethal heredity.

The Potential Outcome

Evolutionists are concerned about this, but they're not surprised. They believe that if we keep those with genetic diseases alive to reproduce, those diseases will keep increasing until the species called Homo sapiens is wiped out. Even if we allow the diseases to take their toll, they fear that mutations will accumulate and turn us into something very different. Dr. James Bonner, a biologist at the California Institute of Technology, puts it this way: "The normal expectation of an animal species such as our own is to arise through mutation, evolution, and selection, and then to die out and be replaced by a species more fitted to the then current environment." Bonner believes that mutations—genetic mistakes in the transmission of heredity—have already formed millions of species of different organisms since life boiled up in the sea over three billion years ago. Most of these species, he thinks, have been wiped out and replaced by hardier organisms.[3]

More optimistic evolutionists hope that humanity can rise above the past, grab the reins of the evolutionary force dragging us down, and create a new humanity through genetic selection. This includes genetic testing of the preborn, abortion of "defectives," transferring genes within the population, and, of course, cloning. Also, if any babies managed to slip through the genetic net, when old enough, they would be required to report to a selected doctor and be sterilized.

This can't be done to members of a free society. However, today prospective parents can be offered genetic testing that can show them the likelihood of their unborn child having specific diseases.

In the next chapter, we'll look in some detail at genetic testing and how it is being used to influence parents in making hard choices about the survival of their children. We'll see how this relates to "scientific breeding," which now must include the very real possibility of human cloning.

9

Genetic Testing
and Scientific "Breeding"

We'll start with some documented cases that came under the genetic microscope a few years back.

Genetic Tragedies

Ten-year-old Monica Anderson's body appears to be turning to bone. A hard crust has grown over her muscles and joints. She is in constant pain, and every breath is a struggle because her rib cage is hardening over her lungs. Dr. Michael Zasloff, a geneticist at the National Institute of Health in Bethesda, Maryland, where Monica is hospitalized, says she might survive only a few more months, or she may live to be sixty. The disease (classified as myositis ossificans progressiva), he adds, is so mysterious that it could stop as quickly as it started.

123

Across the country in San Diego, a little girl just five years old has died of "old age." Doctors at Children's Hospital say Penny Vantine aged at a rate of fifteen to twenty years every twelve months. At her death, her hair was dry and sparse, her face drawn. Her skin was nearly transparent, showing the veins in her forehead. She had glaucoma and cataracts in both eyes. Her circulation was poor, her blood pressure was high, and her fingers were swollen with arthritis. The doctors had no cure for this disease, which they thought came as a result of a disorder in the child's metabolic or endocrine system, probably caused by a genetic defect.

It is the verdict of "no cure" that makes genetic diseases so heartrending. Medical science can only treat the symptoms and help victims live as comfortably as possible. In some instances, control therapies can help people live almost normal lives; but often, little can be done to slow the ravages of mind and body that result from inherited afflictions.

In recent years much research money has gone for developing new drugs and other therapies that can control genetic illnesses. Now, because of new knowledge of genetics, more emphasis is being put on preventing the birth of children with genetic defects. More people are also being trained in genetic counseling. This service has been available at a few university hospitals for several years but was little used because of lack of public awareness. Now, it is likely that most gynecologists recommend it to patients with a high risk for bearing affected children.

Spina bifida is a genetically caused birth defect in which the end of the spinal cord has little or no covering. It causes varying degrees of paralysis of the legs, bowels, and bladder. In most instances, excess spinal fluid also collects in the head, causing hydrocephalus, which, if not treated, results in mental retardation. Each year, twelve thousand children are born with this defect in the United States. The counselor can say that the general risk in the population is one in six hundred, but the risk for recurrence following the birth of a child with an open spine is a foreboding one in twenty-five. Having a previous child with a genetic defect is a frequent reason for seeing a genetic counselor.

A more complicated case is two first cousins who want to marry in a state where such marriages are permissible. The man has a retarded brother with phenylketonuria (PKU). He and his fiancée consult a genetic counselor about the possibility that they might have a child with PKU. The counselor says, "You are wise to be concerned. Couples who are related run a much higher risk than those who are unrelated." Then he writes on his chalkboard:

Frequency of PKU carriers in general population = 1/100
Chance of two carriers marrying = 1/100 x 1/100 =
    1/10,000
Chance of two carriers having PKU child = 1/4
Chance of unrelated couple having PKU child = 1/10,000
    x 1/4 = 1/40,000

Directing his attention to the man, the counselor says, "Because you have a brother with PKU, you run a higher risk no matter whom you marry." Then he writes:

Chance of you being a carrier = 2/3
Chance of you marrying a carrier = 1/100
Chance of PKU child = 2/3 x 1/100 x 1/4 = 1/600

"Now, in the case of you and your fiancée," says the counselor, "the risk is greater." Again he turns to write:

Chance of you being a carrier = 2/3
Chance of cousin being a carrier = 1/4
Chance of PKU child = 2/3 x 1/4 x 1/4 = 1/24

Controversy in Genetic Counseling

Until a few years ago, about all a genetic counselor could do was take a family history and figure probabilities based on Mendel's laws and other related knowledge. Now blood tests can identify the carriers of some diseases, and prenatal tests can tell if an unborn child has certain genetic defects. The prenatal tests are by far the most controversial, since findings frequently influence a couple to have an abortion.

Amniocentesis

For a number of years the most widely administered prenatal test has been amniocentesis. It is usually done between the fourteenth and sixteenth week of pregnancy. A long needle is inserted into the amniotic sac, where the baby is developing, and a sample of amniotic fluid is withdrawn. (There is a 1 to 5 percent risk, depending on the skill of the surgeon, that the needle will injure the fetus or stimulate a spontaneous abortion.) Fetal cells are then taken from the fluid and grown in a tissue culture for seven to twelve weeks. By the time the cells have been examined, the pregnancy is twenty-one to twenty-eight weeks advanced.

Amniocentesis can enable the genetic counselor to tell a couple, with a less than 1 percent error margin, whether or not their baby has Down's syndrome. Without the test, they could be told only their probability of having a Down's syndrome child: one in fifteen hundred for a twenty-five-year-old mother, one in three hundred for a thirty-five-year-old mother, one in twenty-five for a forty-five-year-old mother. Many gynecologists advise pregnant women over thirty-five to have the test for this reason. In some states it is offered free to any pregnant woman past thirty-five.

126

In another situation, a couple is concerned about hemophilia. The bad gene for this disease is passed from a carrier mother to her child on a one in two risk ratio. However, only a son can get the disease, while a daughter can just be a carrier. In this case, the couple has come to the counselor because the pregnant wife has learned that her only brother died at age seven from uncontrolled bleeding. Until reading a magazine article, she had never thought there was any familial connection. Amniocentesis will tell this couple if their unborn child is a boy. If so, a further blood test will indicate if their son has hemophilia. The blood will be checked for a critical clotting substance called Factor VIII. If this element is not present in sufficient amounts, the child's blood will be unable to clot, and he may die of bleeding from the slightest wound.

Before amniocentesis became available, parents could not find out if their unborn child suffered from Down's syndrome, hemophilia, or from 450 other genetic diseases that can now be identified through prenatal testing. Infanticide was out of the question then; so they had to accept the child's affliction. Some parents kept Down's syndrome children. Some had them institutionalized in private or public facilities. Hemophiliacs had to be watched constantly or given injections of synthetic Factor VIII at tremendous cost. With amniocentesis, parents can pretty much know the child's state of health before birth. If the test results are positive for some genetic disease, the possibility of abortion is held out before them.

Genetic Testing and Abortion

In the minds of many people, there is more to consider than the interest of the parents and child. Taxpayers may eventually have a stake in their decision.

The cost of caring for children with severe genetic defects in public institutions is staggering. The time may come when pregnant women will be required to have amniocentesis and, if the test results are positive, to have a forced abortion.

The lead sentence in a recent *USA Today* cover story on genetic testing was: "Genetic testing is changing who gets born in America." *USA Today* researched the state of genetic testing in interviews with doctors, genetic counselors, ethicists, and parents throughout the United States. The reporters "found a booming field in the midst of a string of scientific breakthroughs, but also one torn by wrenching ethical debates." *USA Today* found in 1997 that at least half of the pregnant women who were told their fetuses had serious genetic disorders would decide to have an abortion. Only California, however, keeps exact numbers on how many women have an abortion after learning from a genetic test that their pregnancy has problems. Mark Evans, an expert on fetal diagnosis and treatment at Wayne State University in Detroit, told *USA Today*, "We give good news to most couples." Without the tests, he asserts, many wouldn't risk having children. With the tests, Evans says, "there have been many more babies born as a result of prenatal diagnosis than pregnancies that have been terminated because of it."[1]

Insurance companies are also very interested in genetic testing. Some companies may decide to insert clauses in future policies to protect themselves against claims unless policyholders agree to prenatal testing and abortion if test results prove unfavorable. Geneticists tend to be overwhelmingly in favor of such tests. Dr. Sara C. Finley, codirector of the Laboratory for Medical Genetics at the University of Alabama, Birmingham, agrees with Mark Evans: "Many women with histories of genetic disease would be afraid to have a baby without amniocentesis."[2]

Pro-lifers tend to oppose prenatal genetic testing. One of the most outspoken physicians is Dr. C. Everett Koop, former U.S. surgeon general. Koop is world famous for his surgical skills in saving the lives of newborns with congenital malformations. He teamed with Francis Schaeffer to prepare the book and film series *Whatever Happened to the Human Race?* Koop claims the whole "system" of prenatal testing "is to find out if there is something wrong with the fetus, and if the fetus is defective some parents will decide to abort it. Since I take a high view of life, I see amniocentesis as a search-and-destroy mission."[3]

Many supporters of genetic testing are troubled by couples who abort their child on the basis of its sex. Some doctors refuse the test to couples who admit this is their purpose. But Dr. John C. Fletcher of the National Institute of Health believes doctors cannot prevent patients from having the test for this reason. In 1979 he said, "If you take the position on abortion that the Supreme Court takes, you can't logically and consistently uphold refusing these people."

The battle over abortion has centered around a question of whose rights must prevail, the right of the mother over her body and the new life she carries or the right of the child to live. Those who oppose abortion on demand believe that fetal life is sacred and should take precedence over any inconvenience, short of death or serious injury, to the mother or someone else. Pro-lifers appear to be gaining ground. Some doctors who once championed the right of a woman to have an abortion have changed their minds. Among these is Dr. Bernard N. Nathanson, once one of America's leading abortionists. He is now calling for the outlawing of abortion or for a constitutional amendment protecting life in the womb except when the mother's life is in danger.

Creating a "Better" Human Race?

Many biological scientists think that abortion of "identifiable genetic defectives" will improve the human race as a whole. Dr. Cecil B. Jacobson, chief of the Reproductive Genetics Unit at the George Washington University Hospital in Washington, D.C., is one of the early developers of amniocentesis. He clearly sees it as a tool for fashioning a better future race. He is all for aborting Down's syndrome children, or any other child parents don't want, including a healthy child whose parents prefer the opposite sex. "I just don't recognize any absolutes here," he says. If prenatal testing can be developed to show genetic preconditioning for diseases such as cancer, he would abort these children. "If we could tell what fetuses are going to be afflicted with cancer in their forties or fifties, I would be for aborting them now. That would eliminate some types of cancer forever."[4]

Does that shock you? Well, many believe we can only have a perfect society, a "brave new world," by getting rid of the unfit before they are born. Who will decide the criteria for fitness? Parents are supposed to make those decisions now. Yet recent studies indicate that they are greatly influenced by doctors who do the testing. Dr. John Fletcher followed twenty-five couples of varied social, ethnic, and religious backgrounds who were told their children had genetic diseases. After "counseling," all twenty-five opted for abortion and then were sterilized.

Some Authentically Beneficial Applications

Some genetic defects now identifiable can be modified or controlled before birth to prevent further deterioration in health. Certain enzyme deficiencies can be detected in a follow-up to amniocentesis. One of these deficiencies, methylmalonic acidemia (MMA), is marked by a lack of vitamin B_{12}. If not treated, the child will be mentally and physically retarded at birth. Happily, before the birth of an MMA child, the mother can be given daily injections of B_{12} that will be absorbed by the fetus. After birth, the child can be put on a low-protein diet and given daily doses of B_{12}.

Postnatal medical advances are making more control therapies possible for affected children as well. PKU, for example, can be detected at birth from a pin prick of blood or a drop of urine. The disease is caused by inadequate metabolism of protein. A special diet will usually check the retardation that may otherwise result. Testing for PKU is now required in many states and strongly recommended in others.

Some detectable genetic diseases may require surgery if death is to be averted. One is familial polyposis of the colon. Multiple small benign polyps begin showing up in the colon, although no outward symptoms may occur. This portends cancer of the colon by age forty or fifty. Half of the siblings and children of victims will develop the polyps and eventually have the cancer. The only way to prevent the cancer is to remove all or part of the colon in a colectomy when the polyps are still benign— drastic treatment for a symptomless patient. Dr. Hymie Gordon, a consultant in the Medical Genetics Section of the Mayo Clinic, discovered this disease in a large South African family. He located four hundred "first-degree" relatives (children and siblings) of affected people, half of whom were due to develop the polyps and subsequent cancer. At his recommendation, members of the extended family began coming in for examination and possible colectomies.

The Dangerous Influence of Eugenics

Such treatments do not cheer the scientists bent on "saving" the race from genetic doom. By saving the lives of victims, they say, we are further polluting the population gene pool. If left alone, many of the genetically impaired might not live to bear children. At the very least, the genetic engineers argue, the worst genetic cases should be singled out and sterilized. This gets into eugenics, the science of race improvement. Eugenics (literally "good genes") calls for selective mating to produce well-born children. Southern slave owners before the Civil War sought to improve their slaves by having choice young black men impregnate female slaves of childbearing age. Hitler bred selected young German Aryan men and women to produce "superior" children. Modern agriculturists and cattle raisers practice precision eugenics in improving their crops and herds.

Sir Francis Galton (1822–1911), an Englishman, is called the "father of modern eugenics." A cousin of Charles Darwin, he thought that club feet, curvature of the spine, and a high-arched palate were inherited marks of criminality. Drunkenness, epilepsy, and poverty, he felt, were other evidences of "bad seed." He proposed laws to prevent persons displaying such bad characteristics from mating. He suggested that "healthy" young men and women be given certificates of merit and be encouraged to reproduce. Christians were appalled by his ideas, but many English atheists and agnostics climbed on his bandwagon. Playwright George Bernard Shaw called Galton's proposal the only "religion [that] can save our civilization from the fate that has overtaken all previous civilizations."[5]

Galton's idea fired imaginations in both England and the United States. The number of mental patients sterilized in the United States as a result will probably never be known; most of the incriminating records were destroyed. In a single Kansas home for boys, forty-four "unworthy" young men were castrated on a single occasion. The Indiana Legislature passed a law in 1907 requiring sterilization of "idiots, imbeciles, and the feeble-minded." Similar actions were carried out in other states, with and without legal sanction. In Virginia, relatives of a young "feeble-minded" woman engaged lawyers to defend her rights. They appealed the case to the Supreme Court on grounds that the state courts had denied her equal protection of the law under the Fourteenth Amendment. "Three generations of imbeciles are enough," declared Chief Justice Oliver Wendell Holmes, in supporting the Virginia courts. The young woman was sterilized.

The Nazi horror during World War II stirred public revulsion against human eugenics. Still, the idea for keeping "unfit" people from reproducing did not die. In 1965 a report on criminal patients at Carstairs Maximum Security Hospital in Scotland claimed that 3.5 percent of the male inmates had an extra Y chromosome. The Carstairs study was not taken seriously until a story surfaced that chromosome analysis had revealed Richard Speck, the Chicago mass murderer of nine nurses, to be an XYY. Coincidentally, researchers at Harvard and in Sweden developed special fluorescent stains that made it possible to read some chromosome patterns in cells taken from the inner lining of the cheek. More voices began calling for mass screening of newborn infants to detect criminal tendencies. Before the scheme could be put into action, other scientific studies were published that contradicted the Carstairs results. One researcher reported finding XYYs among a group of businessmen, clergymen, and factory workers. The embarrassed promoters of screening conceded the error and backed away.

About the time the XYY scheme was put to rest, an outcry arose among African Americans in the Civil Rights movement for government funding to research sickle-cell anemia. Support of such research, they said, would indicate how much America was committed to the cause of helping minorities. In the propaganda that followed, the distinction between sickle-cell carriers and victims became blurred. Eight percent of all African Americans were said to be carriers of the trait that "threatens to cripple or kill" (a frequently heard phrase). Newspaper articles and television specials highlighted the complications of a sickle-cell attack: oxygen-starved red blood cells would clump together and clog blood vessels, resulting in painfully swollen joints, damage to kidneys, and lowered resistance to infection. Twenty-nine states and the District of Columbia quickly enacted voluntary screening programs. Some influential politicians demanded that the screening be made compulsory for all African Americans. But the crusade backfired. "Are you a carrier of sickle cell?" became a standard question put to African Americans in job interviews. Airlines stopped hiring them for flying duties on grounds that lowering of oxygen could cause a sickle-cell attack. Some life insurance companies raised rates for sickle-cell carriers.

Scientific research finally cleared the air. Few sickle-cell carriers would actually ever get the disease. Only under conditions of extreme oxygen lack was an attack likely to occur. None of the African athletes at the high-altitude Mexican Olympic Games in 1968 had experienced a sickle-cell "crisis." No documented reports could be found of a single sickle-cell carrier having an attack during an airplane flight. Yet the myth continues today, and some still suffer job discrimination because of false information.

Humanity's record in eugenics is not very good. Still, the race improvers have not given up. Armed with refined lab technology and factual studies of real genetic horrors, genetic engineers are pushing for an expansion of screening and restrictions on marriage for the worst risks. They point out that some laws for eugenics are already on the books: It is illegal for brothers and sisters to marry in all states, and some jurisdictions require sterilization of the severely retarded. Pressure for solutions to genetic problems led Congress to pass the National Genetic Disease Act, creating a special unit within the Department of Health, Education, and Welfare. This unit is providing more financial aid for research, training of genetic counselors, and public education programs concerning genetic disease. Additionally, many states are stepping up genetic screening. All babies born in New York State, for example, must be checked for five inherited diseases.

The Christian's Responsibility

The necessity for education in genetics can readily be seen. We need to be aware of genetic problems in family lines. We need to know how to seek genetic counseling and determine the risk factors in producing children. When we have the facts, we should make responsible decisions about having children. This responsibility goes beyond ourselves. If I have friends who produce a child with cystic fibrosis and they do not understand how the disease runs in a family, I have an obligation to warn them that there is a one in four chance that their next child will also have the malady. Doctors should certainly know enough about the laws of genetics to know when to refer patients for genetic counseling. Pastors should have some awareness for premarital counseling.

I believe God is sovereign and carries out his will among us. Yet I don't think it's right to ignore obvious disease symptoms or flout common rules of health, and then charge the tragedy off to the will of God when the inevitable happens. The businessman who drives up his blood pressure to keep ahead of the pack is not, in my judgment, cooperating with the will of God. Nor is the young wife who knows she carries the gene for hemophilia doing God's will by becoming pregnant and taking the grave risk of having a hemophiliac son or a daughter who is a carrier.

I am personally opposed to amniocentesis when intentions are to abort a child if the presence of a genetic defect is indicated. I don't think either hemophilia or Down's syndrome, which can be revealed by amniocentesis, offer grounds for abortion—certainly not for Christian parents. In any situation, parents should reflect soberly on the findings. If testing indicates that the child will be born seriously deformed, they should probably be prepared to expect a long period of hospitalization after the birth of the child.

A genetics counselor is sometimes put in a predicament concerning how far to go in informing the relatives of a patient. Take tylosis, a rare inherited skin disease easily detected by a mild rash of the palms. On average, 75 percent of the people who have tylosis will eventually develop cancer of the esophagus, which is almost always fatal. The counselor will give his or her patient this information. Should the counselor depend on the patient to tell the relatives that they should be checked for tylosis? Or should the counselor notify them?

What is the counselor's responsibility when a patient has Huntington's chorea—a disease that doesn't become apparent until middle age? Again, should the counselor depend on the patient to inform the family, or should the counselor pass on this information? Should the siblings be told at all, leaving them to live in fear for years to come?

The counselor's job is to inform and not to make recommendations. However, patients are apt to ask, "What do you think I (we) should do?" If the patients are Christians, a Christian counselor can tell them to pray about their problem and perhaps talk to a trusted spiritual counselor, such as their pastor. Still, that may not be enough, even for Christians.

More questions: Should the counselor also get into value judgments about whether a carrier of hemophilia should have children or whether two carriers of the cystic fibrosis gene should marry and reproduce? Does the genetics counselor play God by helping the individual or couple reach a responsible decision about having a child? Do parents play God when they decide?

At this time almost all genetic decisions are a matter of individual choice. As better testing techniques become available, making it possible to identify more defects in the unborn and in carriers, more people will be faced with more choices. The way our nation is moving at present, I foresee that some of these choices will eventually be made by government decree. If enough people buy the fiction that "it takes a village" rather than loving parents to raise a child, I fear that the "village" will be telling many genetic carriers whether they can marry, whom they can marry, and if they can have children. It may reach the point that a couple must get a license to have a child.

We have not seen the whole picture of genetic engineering. Test-tube babies, artificial insemination, cloning, genetic testing, and abortion of the unfit all pale in comparison with genetic engineering designed to reshape human heredity. This will be the subject of the next chapter.

10

Reshaping Heredity

In the summer of 1939 Albert Einstein, perhaps the greatest mind of the twentieth century, wrote President Franklin D. Roosevelt, urging him to investigate the possibility of building an atomic bomb. Einstein, a German Jew who had fled Nazi persecution, was then doing research at Princeton University. Einstein feared Hitler's scientists might develop the bomb first and conquer the world. Spurred by Einstein's letter, Roosevelt authorized a crash project. Two years later a team of scientists, working at the University of Chicago, produced the first man-made chain reaction of atomic energy and unlocked the awesome power of the atom. Upon realizing what they had accomplished, one of the scientists reportedly reflected, "God would not have wanted this."

Many people feel this way about the biorevolution. They think scientists may be displeasing, perhaps even defying, God by developing sperm banks, conceiving babies in laboratories, and devising sophisticated tests that can determine a baby's sex and genetic health before birth. But what about recombinant DNA experiments, through which scientists are striving to reshape heredity itself? The potential for changing hereditary patterns of human life and creating single-sex families through cloning is far more foreboding than artificial insemination and test-tube babies.

The following story illustrates what the recombinant DNA researchers are about. A mountain man came upon a city camper trying to get a bucket of clear water from a muddy stream. The native observed the visitor's futile efforts for a moment, then suggested, "Stranger, if you'd go up and run that hog out of the spring, the water would clear up." The more immediate goal of some DNA research is to clean up the "spring" of life by replacing bad heredity with good heredity, thus eliminating harmful mutations from the human gene pool and creating "ideal" human specimens for cloning. In the process, some scientists think it may be possible to halt and perhaps even reverse the aging process. The long-range intent is to develop superhumans with immunity to all diseases. "Such evolutionary developments," *Time* has suggested, "could well herald the birth of a new, more efficient, and perhaps even superior species. But would it be man?"[1]

The idea of reshaping humanity is being taken seriously by universities, government planners, and the media. One would have to be blind and deaf not to have heard something about DNA. Yet I find that many people are puzzled, especially those who haven't read anything on biology since they were in high school twenty-five years ago. One parent, whose daughter was studying microbiology in college, put it this way: "This DNA code of life I hear her talking about makes about as much sense as the new math homework which I had in school." New math has lost favor in scholastic circles, but I can assure you that DNA is not going to disappear. If it did, we'd all go with it.

The Search for the Secret of Life

Gregor Mendel, you will recall, knew only that traits are passed from generation to generation with mathematical precision. By the time biologists began paying attention to his discovery, they suspected that hereditary information was inside the pairs of tiny threadlike strands in cell nuclei called chromosomes. They noticed that during cell division, the strands always split lengthwise, giving each offspring cell a full share of the genetic material. Thomas Morgan confirmed this with his famous experiments with fruit flies.

By the 1940s biologists were not just suspecting but *assuming* that heredity was in the pairs of genes strung inside the chromosomes. Back in 1871 a Swiss biochemist, Friedrich Miescher, had identified DNA (deoxyribonucleic acid) as being present in cell nuclei. DNA's role in heredity, however, was not understood until 1944, when scientists at the Rockefeller Institute mixed some DNA from genes of pneumonia bacteria with harmless bacteria in a lab dish. The harmless bacteria turned virulent, a demonstration that DNA carried a genetic message. But the scientists couldn't understand how it had happened. One day in 1953 two young scientists were overheard talking loudly in an English pub near their lab at Cambridge University. Someone asked James D. Watson and Francis Crick what they were so excited about. "We have discovered the secret of life," Crick shouted. Crick exaggerated, but their achievement was still stupendous. Crick and Watson had built a Tinkertoy-like model of DNA to explain the code by which the chemical directed the building of life.

141

Their work, based primarily on X-ray diffraction studies by other scientists, won them the Nobel prize and forced biology books to be rewritten.

DNA

Crick and Watson showed that DNA is shaped like a spiral staircase. The "banisters" are composed of long links of sugars and phosphates. The steps between them are made of four different pairs of chemical bases weakly connected at the center. They coded the chemical rungs A (adenine), T (thymine), C (cytosine), and G (guanine). They showed that the sequence of A, T, C, and G could vary widely, providing an almost limitless information-storage system, somewhat like the memory of a computer. Because A always pairs with T, and C with G, one side of the staircase was something of a genetic mirror image of the other. In cell reproduction, they depicted the DNA molecule as unwinding and unzipping down the middle of the staircase, with the pairs A-T and C-G breaking apart at the center. New bases replaced the missing halves of each strand, forming two identical copies of the original staircase. In this way DNA passed on its genetic orders to new cells and future generations.

Watson and Crick's explanation unleashed a deluge of new research projects. Biologists feverishly sought answers to more questions about the chemical nature of life. The large protein molecules called enzymes were identified as the chemical engineers of life. Enzymes are long chains made from the twenty different molecules— the amino acids. How did the DNA staircase order assemble these acids into protein? More studies revealed that at times, when DNA unwinds and unzips, instead of new DNA forming, a complementary messenger RNA (ribonucleic acid) forms to carry the genetic message.

Messenger RNA goes to chemical interpreters (ribosomes) that "read" the building plans. As the code is read, more coworkers (transfer RNA molecules) carry the appropriate amino acids to the ribosomes, where they are hooked together into a protein chain.

How does DNA use a "telegraph" system of only four code letters (A, T, C, G) to select among twenty amino acids for producing complex proteins? George Gamow, a physicist, compared the four bases or rungs of the DNA ladder to different suits in a deck of playing cards. Gamow proposed that DNA's four bases were "drawn" three at a time. Thus 4 x 4 x 4 would yield sixty-four possible combinations. Researchers at the U.S. National Institutes proved that three-letter DNA code messages could call up every one of the twenty amino acids and even "punctuate" directions by marking the "beginning" and "end" of every sentence. This code is universal. The same four letters, used three at a time, specify the variety of protein-building amino acids in the nuclei of the cells of all living things—evidence that the same Architect drew all of the blueprints.

Armed with this new knowledge about the genetic code of life, scientists have since learned to transfer DNA from one species to another. They are recombining heredities at the level of design. We've long heard jokes about breeding across the lines of species, such as, "What do you get when you cross a monkey with a flower?" "A chimpansy!" All joking aside, this doesn't happen in normal reproduction. God so created each "kind" of creature that it is impossible for one kind to breed with another. Ten times in the first chapter of Genesis, the phrase "after their kind" is used. Usually only very similar members of a kind breed together. More rarely, quite different members of a kind may mate, such as lions with tigers or horses with donkeys—but always within the kind.

Recombination of DNA always occurs when a sperm fertilizes an egg. The new recombinant DNA research, however, has started a brand-new chapter by altering the order of creation. Some scientists who don't believe in divine creation are frankly frightened. They're voicing the same fears that are surfacing about cloning: "Don't interfere with nature's basic way of transferring heredity."

Here is what happened.

Reshaping Nature

Scientists worked with Escherichia coli, a rod-shaped, one-celled microscopic bacteria. E. coli, as it is called, has long been one of biological science's most basic research tools. E. coli is a normal resident of the human intestine, where it provides us with some vitamins as "rent" for our providing it a home. It usually does not produce disease, and like other bacteria, it reproduces by simply dividing in two. In the laboratory, E. coli doubles about once every twenty minutes, making it possible to produce millions of cells in a few hours. And, finally, consisting of only one cell makes E. coli a lot easier to study than an organism like a human being who has billions of cells.

Scientists found that E. coli has relatively few genes made of small closed loops of DNA, called plasmids. They discovered that when two bacteria touch each other, a connecting bridge sometimes forms, and a plasmid from one passes into the other. What would happen if plasmids from other bacteria were maneuvered near E. coli? Not only did the bacterium "invite" the visitors "in"; it also began reproducing, doubling, and redoubling, duplicating the recombined DNA each time. This meant that E. coli could serve as a miniature "pharmaceutical factory," developing "bugs" never before seen on earth. DNA could be mixed in the bacterium to form new combinations of bacteria. Using E. coli, General Electric researchers developed a "bug" capable of breaking down a wide variety of hydrocarbons. They suggested a helicopter might drop a quantity on an oil slick and the bugs would multiply and lap up the oil.

During the excitement of achieving DNA recombination, few voices of caution were heard. Most of the talk was about the miraculous new substances recombinant DNA could produce cheaply. Insulin, which is ordinarily taken from the pancreases of animals, was mentioned as one possibility. Another was Factor VIII, the expensive clotting material required by hemophiliacs.

Enthusiasm cooled, though, when some scientists began to speculate that bugs formed from recombinant DNA might turn destructive. Someone wondered what would happen if the GE bacterium able to lap up oil slicks should get into an oil pipeline or the fuel tanks of a plane in flight. Someone else asked about the possibility of creating a killer bacterium in the lab that could escape and multiply millions of times before being missed. It might get into the water supply of a city like New York and infect masses of people. It might be resistant to every remedy known to medical science. It might wipe out humanity.

Scientists talked about such dread "what ifs," but nobody did anything until cancer researcher Robert Pollack heard that Paul Berg, a scientist at Stanford Medical Center, was planning to insert a monkey virus, SV40, into E. coli. Pollack knew this virus had caused tumors when injected into laboratory animals and had made human cells cancerous in lab cultures, even though there was no record of it ever causing cancer in a human being. He called Berg, but the Stanford scientist saw no reason for worry. Berg felt his experiment was extremely significant. He reasoned that if science could understand why SV40 caused cancerous tumors, a big step might be taken toward understanding how cancer develops. After talking with Pollack, Berg discussed the problem with his associates. They expressed concern that E. coli, armed with the tumor-causing virus, might get into somebody's intestines. Berg then agreed to stop his experiment.

Meanwhile, a researcher in San Francisco, Herbert Boyer, found a better way to remove bits of DNA from cells. Back at Stanford, Stanley Cohen, a colleague of Berg's, discovered a versatile new plasmid that could easily pick up a strange gene and pass it off to E. coli. When this news hit the scientific grapevine, researchers from all over the world began calling and writing Cohen for samples. Cohen unveiled his discovery at a meeting of 140 molecular biologists in New Hampshire during the summer of 1973. The scientists sat agape. Here was an easy way to splice any two kinds of DNA together.

Concerns over DNA Research

The possibility that someone could build a new life-destroying bug now seemed much more fearsome. So an investigatory committee of scientists was appointed. They met at the Massachusetts Institute of Technology in April 1974 and agreed to call for a temporary ban on recombinant DNA experiments considered dangerous. This was followed by a think-tank conference of 134 scientists in California plus a battery of lawyers and selected science writers. The scientists presented reports and discussed and argued the risks versus the potential benefits of recombinant DNA research. The lawyers warned that if a destructive bug should get loose and cause damage, the bug's developers could be sued. The scientists decided to continue the ban and to ask the National Institute of Health (NIH) to specify safety levels necessary for continuance of the experiments.

The reporters wrote stories about the possibilities of Andromeda Strain–type bugs escaping and infecting the population, creating a huge stir. A special U.S. Senate committee met to hear evidence. Appearing before the senators, then secretary of the Department of Health, Education, and Welfare (HEW), Joseph Califano, asked Congress to place federal restrictions on the exotic experiments. Alarmed, the NIH moved quickly to establish four levels of physical containment, ranging from P–1 for the lowest risk to P–4 for experiments with cells of higher animals and animal tumor viruses.

A few scientists continued to oppose all recombinant DNA research. Nobel prize winner George Wald, professor of biology at Harvard, persuaded Mayor Alfred Velluci of Cambridge, Massachusetts (where Harvard and MIT are located), to call a meeting of the city council on the subject. The council asked the two schools to halt all research being done at the P–3 level and above while a citizens' review board studied the situation. In February 1977 the council decided that the experiments could continue—but only under stricter safety standards than the government required.

Fearing that all recombinant DNA research might be stopped by an aroused public and that government support funds would be cut off, scientists began trying to develop ways to keep E. coli under control. Two researchers at the University of Alabama developed a gene that makes it virtually impossible for E. coli to survive in a human body. Still, worries continue and opinion is so sharply divided that research scientists at some universities are barely speaking to one another. Many of the fears are justified. If a really dangerous bug is produced, there's a good chance it will escape. We have to reckon with the margin of error that seems to be part of every human project.

Scientists working on recombinant DNA assure us that all safety precautions have been taken, but I'm still a little uneasy. I keep remembering that in 1974 sixty top nuclear scientists rated the chance of a serious nuclear accident about the same as the likelihood of a meteor hitting a major city—one in a million. On March 22 of the following year, fire broke out at Brown's Ferry, Alabama, where the world's largest nuclear generating plant was located. Fortunately no one was hurt, but there could have been a major disaster with thousands of people killed. The plant was closed for three months. What caused it? A workman hunting for a gas leak with a candle!

Since then we've had the terrible nuclear blow-up at Chernobyl in the former Soviet Union. Before that was the scare at Three Mile Island near Harrisburg, Pennsylvania. The accident could have resulted in a meltdown, causing an explosion that would have ruptured the four-feet-thick concrete walls of the containment building and let out deadly radioactive gases. The dust around that accident has not settled yet. Defenders of nuclear power are still accusing the press of ballooning the story just to get headlines.

Research Pushes Ahead

Recombinant DNA research continues at a fast pace. Five large commercial companies, including Standard Oil of Indiana and National Distillers' Corporation, have been heavily involved in seeking new disease-fighting remedies and new ways to grow food. DNA from fruits that resist cold weather might be transplanted into DNA of fruit that is subject to severe damage by frost. Oranges and grapefruit might be grown in Minnesota. "We're talking about billion-dollar possibilities," says

Ronald Cape, chairman of the Cetus Corporation, the world's largest purely genetics firm.[2]

Research sponsored by another firm, Biogen, has succeeded in transferring the gene for human interferon into bacteria that now readily produce the protein. Interferon fights viruses and some types of cancer. Previously, interferon had been extracted from white blood cells through a process that produced only enough to treat six hundred cancer patients a year at a cost of ten thousand dollars each. It was believed that refinements of the new process might reduce costs to as little as ten dollars per treatment.

Researchers at Genentech Corporation have been searching for a human hormone essential to body growth. They believe this could be used to overcome dwarfism. Treatment of one child for this genetic defect has required extraction of hormones from the pituitary glands of fifty human cadavers each year.

For some time there has been talk—just talk—that the aging process might be slowed or even delayed by recombining DNA in cells. Years ago Dr. James Bonner predicted that the "capability [to live two hundred years] will be within our grasp in the next several generations."[3] Dr. F. Medvedev from the Russian Institute of Medical Radiology, however, notes that the older a person becomes, the more likely he or she is to produce a child with a genetic defect. Our life span seems calculated to prevent us from having children after mutations reach a scale harmful to the species. And Dr. F. Marott from the Boston University School of Medicine has theorized that the autoimmune mechanism is switched on with the advance of aging. By the release of antibodies, we devour ourselves. Another theory is that DNA is progressively shut off as we age.

There has been a recent spate of articles suggesting that researchers may be on the brink of discovering how to keep cells from aging. I'm inclined to think that aging is programmed into our cells and little can be done about it. Scripture says, "It is appointed for men to die once and after this comes judgment" (Heb. 9:27). That doesn't mean you must die at a specific, foreordained time. It does mean, though, that death at some point is certain, and there is no way to avoid it, short of the return of Christ and the setting up of his eternal kingdom. The average life span for U.S. males shot up from 53.6 years in 1900, to 70.0 in 1980, to 71.1 in 1990, to 72.1 in 1993. For females the jump was from 54.6 in 1920 to 78.9 in 1993. The life span for males increased only 1.1 from 1980 to 1990, while expectancy for women rose from 70.0 to 71.8 during that period. From 1950 to 1993 the average life span for U.S. men increased only 6.5 years. Female expectancy jumped 7.8. Some scientists think that under optimum conditions we will go no higher than 110. This is average, of course.[4] The only people we know to have lived extremely long lives are the patriarchs of Genesis 5. These men, I think, had special genetic constitutions selected by God and were not typical of the human species.

Recorded human life spans shortened dramatically after the Flood, even though Abraham lived to be 175. I suspect that the wives of Noah's sons were from the short-lived varieties of humans. Their genes were mixed with the long-lived variety (Noah's sons) and life expectancy began dropping. Later the Bible speaks of seventy as the age (in round numbers) to which a person could expect to live. It's interesting that we're still only a few years above that mark in this day of scientific genius.

I'm not optimistic that DNA researchers will find a way to significantly increase maximum ages. All of us need to prepare to die. I do predict that many new substances will be developed through recombinant DNA research for treatment and, in a few instances, for control of a number of troublesome diseases. Whether science will be able to deal with these diseases at their hereditary source is another matter. Some optimistic researchers say that we will learn how to manipulate the control mechanisms in our DNA. It will then be possible, they say, to take some of a person's DNA and grow spare organs in the lab, ready when old ones wear out. The real dreamers think that human anatomy can be reshaped. The time will come, they say, when we can have extra thumbs, a larger brain, longer or shorter legs, and perhaps X-ray eyes and super-hearing. It is even speculated that the twenty-fifth-century person will consist of a central brain connected to a variety of indestructible limbs and sense organs that may travel as he or she fancies.

There are also predictions that it will one day be possible to eradicate "undesirable" attitudes and feelings. *Science* magazine reported in the fall of 1996 that new evidence of a worry gene had been uncovered. Worrywarts, says Edward M. Hallowell of Harvard Medical School, "can, in part, thank genetics and brain chemistry for a tendency to fret. The ultimate goal is human perfection."[5] I'm not saying that human beings will become superbeings from new research or that God will permit near physical perfection to come to pass. All I am saying is that biologists have already accomplished many feats that were once thought impossible and that others are on the horizon.

Our Response

Having a super mind and body—if the means are discovered—will not necessarily spell happiness. I have strong, mixed feelings about the potential benefits to be gained from earthshaking biological "improvements" to the human race. Maybe God will decide that we have become too proud and self-sufficient and will allow a one-celled monster to escape a lab and devastate a city just to bring us off our thrones. The ten plagues of Exodus surely brought Pharaoh down from his pinnacle of pride and rebellion.

Some scientists speak of evolution in terms of God. They warn their colleagues against tinkering with the delicate mechanisms of evolution. Others see evolution as an enemy to be conquered. "Evolution brought us typhoid and yellow fever epidemics and outbreaks of polio," they say. "Evolution will kill us all if we don't get control of our destiny."

What can we say as Christians who believe in special creation? We can say God is in control. Nothing ever happens that surprises him. We can say there is meaning in creation. God created life "after its kind" and divided the kinds for good reasons. He created the process by which we are conceived, born, and live until we are cut down by the corruption brought by sin. He gave all of life a beautiful balance. We need to be careful about how we upset this balance and question the wisdom of God.

Apart from such philosophizing, we must face the real questions that confront us from unexpected new discoveries in genetics and from formulas for reshaping our heredity. What are the permissible boundaries for experimentation with the basic substance of life? When do the risks become greater than the potential benefits? What is perfection? Is it the absence of discomfort and pain? Is it an Arnold Schwarzenegger physique? Could the absence of disease ensure a perfect world? Can selfishness and hostility be genetically programmed out of us and replaced with unselfishness and love? Who will decide which traits are worth keeping and which should be altered or eliminated? How should Christians respond to the challenges of science in the biorevolution?

I do not have the final answers to these questions. But God knows. If we turn to him, he will show us the way.

11

Where Do We Go from Here?

Twenty years ago a professor was going around the country telling audiences, "If you're alive twenty years from now, you'll live forever." His name was F. M. Esfandiary, and he was then teaching courses in futurology at New York's New School for Social Research. In 1979 he predicted that by the year 2000 science would have conquered disease, pain, and even death. By 1990, he predicted, there would be a microcomputer that could be implanted in our bodies for constant monitoring of every vital organ. If your heart started acting up, for example, the computer would tell you in plenty of time to get expert medical attention. The aging breakthrough, Professor Esfandiary said, will come through genetics. Science will find a way to alter DNA so that the body will keep reproducing healthy cells forever. (I know of no qualified geneticist who has made a similar claim.) Because Esfandiary claimed to be a scientist and a professor, the newspapers picked up his prediction, and many believed him. He was telling them what they wanted to hear. Then we have the recent moneymaking *Jurassic Park* and *Mimic*, with their stories of DNA manipulation. Between the snake-oil claims of people like Esfandiary and Hollywood fantasies, many people think the biorevolution is just a big put-on. "You don't really think they're going to clone a human, do you?" "This DNA business is a lot of double-talk to confuse the bureaucrats and keep the grant money rolling into the universities." So some people say. The actual accomplishments in biological science are about halfway between the cynicism and fantasies, though. The ultimate feats, which science-fiction writers imagine, haven't happened yet, but recent developments are still astounding.

At times all of us would like to turn back the clock on things that shock and disturb us. But we can't. "Time marches on," and there is no turning back. Michael Dukakis, former governor of Massachusetts and candidate for the presidency, observed in *Time* magazine: "Genetic manipulation to create new forms of life places biologists at a threshold similar to that which physicists reached when they split the atom. I think it is fair to say that the genie is out of the bottle."[1] Added Robert Sinsheimer, chairman of Cal Tech's biology department: "Biologists have become, without wanting it, the custodians of great and terrible power. It is idle to pretend otherwise."[2]

Research is going to continue. Government funding may not go on forever, forcing university labs to cut staff, but the top scientists will easily find jobs with private corporations and foundations. The trend in employment of scientists is already moving in this direction. And the warnings will keep coming from people in the know.

Dr. George Wald predicts that the results of recombinant DNA work

> will be essentially new organisms, self-perpetuating, and hence permanent. Once created, they cannot be recalled. . . . It is all too big, and is happening too fast. . . . It presents probably the largest ethical problem that science has ever had to face. Our morality up to now has been to go ahead without restriction to learn all that we can about nature. Restructuring nature was not part of the bargain; neither was telling scientists not to venture further in certain directions. That comes hard.[3]

Wald wants strong federal action to bring all research under control. He fears "for the future of science as we have known it, for humankind, for life on the Earth."[4] Princeton's Paul Ramsey warned that genetic engineering "could cause the genetic death God once promised and by His mercy withheld so that His creatures, despite having sought to lay hold of godhood, might still live and perform a limited, creaturely service of life."[5]

There is great concern over what genetic tinkering with heredity, cloning, and abortion of "experimental" or "defective" children will do to the family. To quote again Dr. Leon Kass: "The family is rapidly becoming the only institution in an increasingly impersonal world where each person is loved not for what he does or makes, but simply because he is. Can our humanity survive [the family's] destruction?"[6]

The bitterest arguments come over aborting unborn children who have tested positive for defects. Joseph Fletcher holds that there is a moral obligation to abort children who do not measure up to a certain quality of life. He also says defective babies and the terminally ill should not be allowed to live if they do not meet a set of "positive human criteria." For Fletcher, an I.Q. of forty is a "questionable person," and below twenty, "not a person." Fletcher has further asserted that people with genetically defective pedigrees do not have a right to reproduce. "Our gonads and gametes are not private possessions," he told the second National Symposium on Law and Genetics held in Boston.[7]

C. Everett Koop, who accepts the Bible as the final authority, puts a different value on "imperfect" children. Says the former surgeon general, who has performed surgery on many little ones born with defects:

> When God spoke to Moses at the burning bush, he said, "Who makes man dumb or deaf, or seeing or blind? Is it not I, the Lord?" Whether you or I like it or not, God makes the perfect and also what we would call the imperfect. This is in His sovereign plan. I don't think I should say that God made a mistake and therefore I am going to get rid of this child.[8]

God's idea of perfection is different from ours. Modern people are body worshipers. You need go no further than a television set or a newsstand to recognize this. God, though, looks at the inner person. I believe that in God's sight, a sweet, gentle Down's syndrome boy is more beautiful than a spoiled, mean-tempered Olympic star.

How Should Christians View Science and the Changes It Brings?

We seem to have two extreme views of science. One view deifies it as the potential savior of all humankind. The other places it on a level with demonism. Some Christians I know automatically assume science is out to destroy their faith, immediately becoming defensive whenever a scientist begins talking about origins. I think that deep in their hearts these people suspect that one can't be intellectually honest and still accept the Genesis record. We can be grateful, however, that an increasing number of scientists are showing us that we can keep our minds in gear and still believe the Bible.

Science is a powerful method for exploring and amassing knowledge about God's creation. This is in the mandate of Genesis 1:28, where God commanded humanity to "subdue" the earth. The earth was made for us to control and preserve in ways that honor God, and science gives us the tools for this. Most of the scholars from past centuries, who laid the foundations for modern astronomy, physics, and biology, believed this. Sir Isaac Newton, the "grandfather" of the rocketry that sent a man to the moon, was also an eminent theologian. The Bible and Christian faith were integral to the achievements of such great scientists as Pascal and Galileo, who were believers first and scientists second. Are you surprised to find Galileo on the list? He was a believer. He just happened to be way ahead of religious dogmatists who had forgotten that God created the heavens as well as the earth.

Three Philosophies behind Science

Science is one thing; philosophy is another. Our philosophy determines what we think about the values and purposes of science and how we use it. Not only that, but the conclusions to which scientists come are often heavily colored by their philosophy.

Let's get down to basics. You can divide scientists into mechanists, theists, and agnostics. Mechanists say that if we knew everything about human beings, we would be totally explainable in terms of physics and chemistry. To put it another way: We are merely matter, and that is all that matters. Theists say science can never completely explain human beings, because the totality of humanity

157

transcends the realm of knowledge available to us. In other words: We are much more than matter, and in the long run, that is what really matters. Theists not only are convinced that humans are more than atoms, they also believe in some form of deity. Agnostics try to operate on the assumption that humans cannot know whether they are machine or soul. This puts the agnostics in a tortured dilemma. Take the decision to destroy an unborn child who has a genetic defect. From the mechanist perspective, the child is only an imperfect machine in the making. From the theist's viewpoint, the child is something above and beyond a machine. Is that "something above" worth saving? If agnostics proceed as mechanists would, they will always wonder if they made a tragic mistake.

The true mechanist will not worry about the consequences of destroying a child who doesn't measure up to some health standard. For that person, there is only a difference in degree, not a difference in kind, between a human and an insect. Fortunately, most mechanists aren't as extreme as that. Some agonize over decisions to abort children. Physical life is sometimes more precious to them than to believers. After all, the flesh and their earthly span of mortality are all they have.

The present generation of mechanists has no real hope of achieving immortality for themselves, despite the foolish talk by those who tread in the footsteps of the few who think science can make it possible for humans to live forever. If this could happen, think of the population problem and the drain on energy resources that would result! "Oh, but humans will find a way to populate other planets," the dreamers say. Ridiculous. We would have to take our environment with us. None of the other planets in our solar system can support human life in their native state. To colonize Mars, or any other planet, buildings would have to be constructed to hold Earth's atmosphere.

Today's mechanists can only hope for a better life for future generations. So they talk grandly of correcting genetic mutants and eliminating disease. Yet to improve life, they realize that they must destroy "inferior" life. How are they going to make the tough decisions about who will live and who will die? How will they ever agree among themselves? Scientists are not a special breed. They are no better or worse than other people. Some hold petty jealousies. Most compete for honors. Some will stab a colleague in the back. Some cheat on income tax.

If it's wrong to worship scientists, it's also wrong to make them monsters of evil. I've already said that some of my fellow Christians (including ministers) have been guilty of the latter. A minister should inform his or her congregation about creationism, evolutionism, mechanism, theism, genetic engineering, and cloning as each relates to the sacredness of human life and individual rights. Unless he or she is well-read in the field or has had scientific training, he or she should let Bible-believing scientists explain the technicalities. There are now hundreds of scientists in the Creation Research Society, an organization whose membership consists of those who find no conflict between the Genesis record of creation and the facts of science. Perhaps the most important work being done today by these scientists is the development of comprehensive concepts about the history of the universe, of the earth, and of life, all of which are faithful both to the teaching of Scripture and scientific observation.[9]

We Christians need to get moving. At this moment, the mechanists are pretty much in command and moving full speed ahead with education programs and social legislation. Occasionally, they go too far and are forced to take a step backward. The Macmillan Company, which has a big slice of the market in public school science texts, once offered in their educational catalog human embryos embedded in plastic. Grisly? You bet. We dissect fetal pigs in my biology lab. Nothing wrong with that, since these animals contribute to our understanding of our own bodies and would otherwise be discarded. But if American society keeps going the way we're heading now, freshmen of the future will probably be dissecting fetal humans "grown" for this purpose. If a human is just another animal, there is no reason why this shouldn't happen.

The Influence of Secular Humanism

I'm both pessimistic and optimistic about the future. I'm pessimistic about ominous trends in public education. Drug use, violence, and sexual immorality worry me, of course, but what concerns me more is the pervasiveness of secular humanism, which recognizes no moral absolutes and sees everything as relative and conditioned by culture. The abortion of unwanted children, genetically diseased or not, is made a personal choice for which one must answer only to the forces within culture. Relativists say homosexuality is a legitimate lifestyle and demands respect. To them, sexuality, like abortion, is only a matter of choice. "Politically correct" gays insist that homosexuality is genetic in origin. But no "gay" gene has yet been discovered, and it is obvious that culture and environment are major determinants in sex roles.

Secular humanism is blatantly promoted in textbooks and classrooms in public and private colleges and universities and in many denominational schools as well. Here's an example from a highly regarded book used in some colleges. The excerpt is from a chapter titled "Morality and Communicational Process" by Dr. Harley C. Shands, chairman of the Department of Psychiatry, Roosevelt Hospital, New York City. Dr. Shands is one of twenty-five authors of the study. Under the topic "Man,

Language and Culture," Dr. Shands suggests that human behavior is reprehensible only within one's cultural group.

All of us are, always and unavoidably, in the position of David, to whom it was necessary for Nathan to say, "Thou art the man" with reference to [his] sin with Bathsheba, and in that of Oedipus, to whom it was necessary for Teresias to convey the same message. In both instances, the power of the consensus can be seen in the ensuing behavior of the rules: Solomon [David], more restrained and more routinely offending, was satisfied to repent; while Oedipus, appalled at the consensually abhorrent crime of incest, was moved to blind himself as his wife-mother Jocasta killed herself.

These mythological examples attest to a truth daily evident in our courtrooms. The defendant is not guilty until said to be guilty: "guilt" is not an absolute moral judgment, it is a relativistic consensual judgment. . . .

In affairs of morality and ethical questions, we find the background of relevance that of relativity and consensus: a culture is primarily defined in its own terms. A culture is analogous to a dictionary—each word (or member) is defined in terms of their words (or members) which only have value in a system including all. This means, in turn, that objectivity is a myth, and perhaps a most dangerous and even lethal myth. In human affairs, there is no such thing as objectivity, and therefore no such thing as a universal system of morality or ethics.[10]

Did you get this learned man's point? Nothing, nothing is absolute. Everything is cultural, even incest and suicide. A couple of other things should be noted. One, he classifies the biblical story with the Greek drama as myth. You see this quite often in high school literature and social science texts. Well, to some extent Dr. Shands's biblical story is a myth, for it was David, not Solomon, who committed adultery with Bathsheba. Solomon was their son. This is not an isolated case. Cul-

161

tural relativism is also the norm in secular higher education. This base for morality is wholly human-centered, not God-centered. Morality comes from within, not without. Everything is up to humanity. Jacques Monod, a philosopher guru among secular humanists and a Nobel prize winner in 1965, calls humans only a product of chance genetic mutation. Monod concludes: "Man knows at last that he is alone in the universe's unfeeling immensity, out of which he emerged only by chance. His destiny is nowhere spelled out, nor is his duty."[11] The secular humanistic philosophy is also present in modern literature, television, movies, and on the Internet. These are some of the moral-ethical values that appear frequently:

1. The quality of personal experience holds priority over one's role as a member of a family, employee and/or employer, and citizen of the community and country. One's first obligation is to oneself.
2. Morality is developed from within one's own mind and experiences and not from institutions— church, state, school, corporate business, which are chained to the past and are no longer relevant to present life.
3. Moral ties and loyalties are linked only to one's immediate group.
4. Sex is only a physical experience and has no moral or ethical relevance beyond emotional and physical self-fulfillment, which may or may not be linked with affection for another. Expressions of sexuality, whether heterosexual or homosexual, are contingent upon style and preference.
5. Honesty is virtuous only within the intimate group. It is all right to rip off outsiders, especially uncaring institutions.
6. There is no purpose or meaning beyond this life.[12]

No moral consensus exists today on absolute rules, such as the Ten Commandments. The worldview of most educators, communicators, government bureaucrats, and many denominational church functionaries can be summed up this way: Do it. Experiment. Be free. Live, live, live. You are all you have. You only go around once, so make the most of life while you have it. This philosophy blossomed in the 1960s and swept the country in the 1970s. It is still tearing apart our social structure, confusing college students, breaking up marriages, and alienating members of families from one another. There is no place here for imperfect children or any other misfits, since they will only tie the "beautiful people" down. Of course, if caring for a handicapped child or an aged parent is something you enjoy, then go ahead and do it.

This is the exact opposite of the Christian lifestyle Jesus asked his disciples to follow: giving instead of getting, serving instead of seeking self-satisfaction, putting the interests of others before your own, and seeking to please and honor God more than anything else. "Whoever wishes to save his life shall lose it," Jesus said, "but whoever loses his life for My sake, he is the one who will save it" (Luke 9:24). Witness the unhappy multitudes who drag from one weary experience to another, seeking a new "high." Witness the depression and suicides of rock stars and movie idols and deaths from drug overdoses.

What does this have to do with the moral problems posed by the biorevolution? Everything. If we answer to no one but ourselves, then science must seek to make the perfect human through laboratory breeding and cloning. If unborn babies cannot contribute to the self-fulfillment of the mother and others, then abort them, whether they happen to have physical limitations or to be of the "wrong" sex. If newborn babies reveal a low "quality" of life, then allow them to die by withholding treatment.

Before the Supreme Court authorized abortion on demand in 1973, C. Everett Koop predicted that infanticide would follow after a million abortions a year in the United States. We've surpassed that figure, and Koop's prediction has become a reality. He attests that infanticide is practiced widely in this country today, that is, the deliberate killing by active or passive means of a child who has been born. Pediatrician Raymond Duff reported in the October 25, 1973, issue of the *New England Journal of Medicine* that physicians at the Yale–New Haven Hospital withheld treatment from a group of sick babies born with physical defects. Fourteen percent died as a result. This is certainly not happening in every hospital today, but it is happening in some institutions of "mercy" in the United States. Indeed, the "partial birth abortion," supported by President Clinton and other leaders, misses being clear infanticide by mere seconds.

Dr. Jack Kevorkian, the celebrated suicide doctor, fits right into this line of thinking. He claims only to be relieving the terminally ill of their pain. Yet he has "helped" some end their earthly lives who were not terminally ill. The suicide binge is worse in Holland, where some doctors have pulled the plug without patient consent.

What Should Christians Do?

An increasing number of people seem to believe that whatever makes you feel happy and fulfilled is okay. In the face of all this, what are we Christians to do? We must begin with our children at home. We must teach by word and example that life is sacred, that God is real. And we will have to take an active role in limiting what they are exposed to, especially via television and the Internet. As our children get older, they will inevitably be exposed to the values of the world. But if they have a clear and firm foundation of faith, they will be able to choose responsibly.

Regarding science specifically, I do not believe we need to start a war against biological research. On the contrary, I propose that we encourage more Christian young people to enter scientific careers, especially science teaching. We have a responsibility to our community and our country in regard to the direction in which the biorevolution is going.

You're against human cloning? So am I. Cloning to improve agriculture and to improve the breed of domestic animals is valid, but cloning humans is one area of research that should be left strictly alone.

Is there some standard that we can use to evaluate all of these new reproductive alternatives? Even though the Bible doesn't address these techniques specifically, I believe that it gives us what we need. The Genesis record reveals God's plan for human reproduction: one husband and one wife. Any deviation from that plan is just that, a deviation to be avoided. On that basis we can accept AIH while opposing AID. Artificial insemination with the husband's semen is a medical technique to allow a married couple to produce their own children. But artificial insemination with a donor's sperm introduces a third person to the marriage relationship and brings with it psychological and sociological burdens that are inappropriate. And, of course, artificial insemination for lesbians should be opposed. Host motherhood likewise breaks the intimacy of the marriage bond, and I think there should be restrictions against a woman bearing a baby for another. Too many problems are involved, both biblical and psychological.

Recombinant DNA research, as I discussed in the previous chapter, is a mixed bag for me. It holds great promise for curing some genetic diseases. It also has vast potential for wrong use. Proper safety precautions must be observed to prevent a dangerous "bug" from escaping and infecting people. I don't like government regulations any more than the next person, but somebody has to set guidelines, and I'm not in favor of researchers policing themselves. What worries me more is the very real possibility of attempts to create a master race through genetic engineering, of which cloning is a part. I'm going to continue to be concerned and watchful.

Prenatal testing has become all the rage in the closing years of this millennium. Defects and diseases that may be uncovered include extra or missing chromosomes, genetic maladies caused by mutations of particular genes (cystic fibrosis and Tay-Sachs disease are examples), and neural tube defects that may indicate anencephaly (an improperly formed brain), spina bifida, and other defects.

I would not condemn married couples who have such tests. In some instances, one—or in rare cases both—of the spouses has a crippling, life-shortening disease that runs in their family. A test before pregnancy may show the chances of the couple having a child with the same or a similar disorder. When the wife is already pregnant, knowledge of disease can guide them in making future plans. Many couples, however, choose to have an abortion on learning of a birth defect, and this I oppose.

Experimentation on humans is another very serious matter. The record shows that scientists can't always be trusted to get proper consent. The possibility of research on live embryos and fetuses for the advance of science makes my blood run cold.

I'm opposed to amniocentesis when abortion is an option. I recognize that some Christian professionals will differ with me here, but I must agree with Dr. Koop that amniocentesis is usually a "search and destroy" mission. Except in unusual situations, a woman will have this test planning to have an abortion if the results turn out wrong. On the other hand, amniocentesis that can lead to prenatal treatment is an acceptable option. There is an increasing number of conditions that can be diagnosed this way, providing the unborn baby with options for treatment that can prevent or alleviate a condition. In the future we may be faced with some touchy legislation about amniocentesis and certain other forms of testing. Unless we get a constitutional amendment forbidding abortion on demand or the Supreme Court reverses the 1973 decision on abortion (which I don't think is likely), we may soon be forced to abort babies with genetic defects. Actively opposing abortion and some forms of testing frightens some people. They say we're trying to impose our moral views on everybody. But consider the moral legislation already on the books. We already have laws against murder, rape, theft, perjury, etc. We need laws to protect the unborn and to keep scientific research within moral bounds. I'm not calling for an inquisition. I'm just in favor of protecting the helpless.

We Christians cannot dodge our responsibilities as citizens in this "brave new world" the biorevolution has brought us. Genetics is something most of us know little about. Some of us have displayed our ignorance in the past by labeling certain diseases as "judgments" from God. Then science comes along and discovers a genetic cause. We can get into trouble by drawing inferences from Scripture that aren't there.

Rather, we need to become informed about basic genetics. Pastors who do premarital counseling should be aware of genetic problems and should find out if any exist in the families of engaged couples. When trouble is suspected, couples should be referred to trained genetic counselors. I say this because genetic risks are not always easy to determine. There are too many exceptions to every generality, and working out the risks in-

167

volves mathematical procedures that may be difficult for the untrained person.

What should Christians do if they learn they have bad genes? I can't give a pat answer here. On some moral questions, Scripture does not speak clearly, and this is one of them. The command to "be fruitful and multiply" was given to Adam and Eve, who were in perfect genetic health. It couldn't apply to every future adult. Many people are sterile and can't have children.

I do believe that all children are a gift from the Lord. But when I talk to my students about family planning, I ask: "Do you have the right to accept the gift of a child if you're not prepared to care for it? Does the potential to produce a child authorize you to go ahead when you can't feed an extra mouth?" These questions likewise apply to a couple who face a good chance of having a baby with few prospects for a healthy life. Consider a couple contemplating parenthood who learn through a genetic test that they have one chance in four of having a child with cystic fibrosis. I would tell that couple: "Maybe God doesn't want you to have a child. Maybe he wants to do something special with your childlessness." But isn't the ability to reproduce an indication that doing so is a God-given right? Not necessarily. There are times, as I mentioned before, when we should consider not having children or at least postponing pregnancy until a clearer genetic path is found. Some couples should probably consider sterilization.

How should a Christian couple respond to the birth of a child with serious health problems? We've been talking about genetic risks, yet every couple faces the possibility of having a child that is less than normal. First, recognize that this child, too, is a gift from God.

What non-Christians see only as a tragic mistake, God can make a blessing. I can't explain all the suffering caused by genetic diseases. I can only say that, in his infinite wisdom, God has a purpose in the birth of each child. It is through suffering that our faith is purified and that we are taught to love in deeper ways than we would otherwise know. Sometimes the teacher God appoints for us is a little child born with severe physical or mental handicaps.

There is a mystery here that should cause us to reflect deeply, very deeply, on some of the motivations and intended consequences of the biorevolution. Where will it lead us? Where is God's guidance? Where is the fear of the Lord that Proverbs says is "the beginning of knowledge" (Prov. 1:7)?

The cloning of human beings is now a very real possibility. Should this happen, our planet will not go up in a nuclear cloud. The event will be a sensation. The media will have a circus. A round of film extravaganzas will hit the television and theater screens. Many who have been opposed to cloning will predict the direst of consequences. Champions of cloning will herald the accomplishment as the dawning of a new age—an age when we will learn to perfect ourselves.

Human knowledge is a precious resource. Used wisely, the knowledge of genetics and heredity will protect and enhance the dignity and sanctity of human life. But let humans tamper with God's order of life for their own ends, and they place themselves in grave danger. The day of reckoning will come.

Notes

Chapter 1: A Brave New World

1. David Rorvik, Publisher's Note, *In His Image* (Philadelphia: J. B. Lippincott Co., 1978), p. 5.

2. "Cloned in USA," *St. Louis Post-Dispatch*, 13 March 1997, p. 4A.

3. Sharon Begley, "Little Lamb, Who Made Thee?" *Newsweek* (10 March 1997), pp. 53–60. "Will There Be Another You?" *Time* (10 March 1997), pp. 60–73. "The World after Cloning," *U.S. News & World Report* (10 March 1997), pp. 59–63.

4. For example, "Science News of the Week: Ewe Again? Cloning from Adult DNA," *Science News* (1 March 1997), p. 132.

5. *Business Magazine* (10 March 1997), pp. 79–86.

6. "Theology after Dolly," *Christian Century* (19–26 March 1997), p. 285.

7. *World* (8 March 1997).

8. Patricia Wilson, "Dolly Scientists Reject Human Cloning," Reuters News Service, 27 June 1997.

9. James D. Watson, "Moving toward the Clonal Man—Is This What We Want?" *Atlantic Monthly* (May 1971).

10. Tim Poor, "Ban Human Cloning Research, Bond Says," *St. Louis Post-Dispatch*, 26 February 1997, p. 9A.

11. Joseph Fletcher, "Ethical Aspects of Genetic Controls," *New England Journal of Medicine* 285 (1971), pp. 776–83.

12. Aldous Huxley, *Brave New World*, with new Foreword (Catchogue, N.Y.: Buccaneer Books, 1932, 1946), p. xiv.

13. Paul Ramsey, *Fabricated Man* (New Haven and London: Yale University Press, 1970), pp. 104, 137.

14. "Oldest Person," *St. Louis Post-Dispatch*, 7 August 1997, p. 2.

171

Chapter 2: In *Whose* Image?

1. Maggie Gallagher, "The Need for Laws to Control Cloning," *St. Louis Post-Dispatch*, 5 March 1997, p. 7B.

2. Ian C. Wilson and John C. Reece in *Archives of General Psychiatry*, October 1964, as cited by Rorvik, *In His Image*, pp. 88–89.

3. Steve Jones, *The Language of Genes* (New York: Doubleday, Anchor Books, 1994), p. 189.

4. Ibid., p. 191.

5. The report on the calf cloning was from Internet site http://www.merc.com/stories/cgi/story.cgi?id=4331281–7f8).

6. "Cloned in USA," p. 4A.

7. "Wisconsin Company Announces Cloning of Calf," *Science News* (7 August 1997).

8. *Science News* (17 February 1979).

9. *Family Weekly* (4 March 1979).

10. Rorvik, *In His Image*, pp. 40–41.

11. Emily and Per D'Aulaire, "Clones: Will There Be Carbon Copy People?" *Reader's Digest* (March 1979), pp. 95–98.

12. Robert T. Francoeur, *Utopian Motherhood* (Garden City, N.Y.: Doubleday, 1970).

13. 19 August 1997, e-mail from srtscot@dial.pipex.com from Scotland.

14. Ibid.

15. Announcement from the White House, March 1997.

16. "Cloned in USA," p. 4A.

17. *Pronouncement* (27 May 1997).

Chapter 3: Will Clones Have Souls?

1. I say "substantial" because there is a possibility of mutations occurring in body (somatic) cells after conception, or in this case, after cloning.

2.. "Statement on Human Cloning," *Light* (July–August 1997), p. 2.

3. *Nature* (8 January 1979).

4. James C. Hefley, *Scientists Who Believe* (Elgin, Ill.: David C. Cook, 1963), p. 42.

5. Quoted in Bob Allen, "Sheep Cloning: A Baaaffling Question," *Baptist Standard* (5 March 1997), p. 1.

Chapter 4: Should Humans Be Cloned?

1. "Detour 39210," *Newton County Times*, 27 February 1997, p. 12.

2. Ira Levin, *The Boys from Brazil* (New York: Random House, 1976), p. 253.

3. Ellen Goodman, "The Shepherds and the Sheep," *Boston Globe*, reprinted in the *St. Louis Post-Dispatch*, 27 February 1997, p. 7B.

4. Ramsey, *Fabricated Man*, p. 71.

5. Ibid., p. 72.

6. William Allen, "Litter Control," *St. Louis Post-Dispatch*, 21 August 1997, p. 15A.

Chapter 5: Cloning Is Not God's Way

1. Donald MacKay, *The Clockwork Image* (Downers Grove, Ill.: InterVarsity, 1974), p. 57.

2. Quoted in James C. Hefley, *Life in the Balance* (Wheaton: Victor, 1980), p. 30.

3. Ibid.

4. Ibid.

5. "Why There Is No Substitute for Parents," *Imprimis* newsletter, Hillsdale College, Hillsdale, Michigan (June 1997), p. 1.

6. James C. Hefley, *Dictionary of Illustrations* (Grand Rapids: Zondervan, 1971), p. 127.

7. Ramsey, *Fabricated Man*, p. 89.

8. *Washington Post*, 30 September 1967, quoted in Ramsey, *Fabricated Man*, p. 102. Also cited in Rorvik, *In His Image*, p. 62.

9. Quoted in Rorvik, *In His Image*, p. 61.

10. "ERLC Cites Flaws in Ban on Federal Cloning Funds," Baptist Press Service in *Word & Way* (14 August 1997), p. 16.

Chapter 6: Test-Tube Babies

1. Rorvik, *In His Image*, pp. 56–60.

2. Allen R. Utke, *Bio-Babel* (Atlanta: John Knox Press, 1978), p. 35.

3. Dietrich Bonhoeffer, *Ethics*, trans. N. H. Smith (New York: Macmillan, 1955), p. 131.

4. Joseph Fletcher, *Morals and Medicine* (Boston: Beacon Press, 1960), pp. 150–51.

5. Howard Brody, *Ethical Decisions in Medicine* (Boston: Little, Brown, 1976), appendix IV.

6. Ibid., pp. 156–58.

7. Ibid., p. 159.

8. Kenneth Guentert, "Will Your Grandchild Be a Test-Tube Baby?" *U.S. Catholic* (June 1977).

Chapter 7: Artificial Insemination

1. Personal interview with James C. Hefley.

2. *Washington Post*, 3 February 1980, quoted in *World*, 8 March 1997.

3. Dr. Laurence E. Karp, *Genetic Engineering: Threat or Promise* (Chicago: Nelson-Hall, 1976), pp. 137, 158.

4. James and Marti Hefley, "Babies in Question," *Today's Health* (August 1970).

5. Ibid.

6. Norman Anderson, *Issues of Life and Death* (Downers Grove, Ill.: InterVarsity, 1974), pp. 49–50.

7. Harman Smith, *Ethics and the New Medicine* (Nashville: Abingdon Press, 1970), pp. 72–73.

8. "Babies Made to Order: How to Stir a Tempest in a Test Tube," *Chicago Daily News*, 20 September 1974.

Chapter 8: The Heredity Trap

1. Kim Painter, "How Much Do You Want to Know?" *USA Today*, 15 August 1997, p. 2.

2. Ibid.

3. Robert Cooke, *Improving on Nature* (New York: New York Times Book Co., 1977), p. 15.

Chapter 9: Genetic Testing and Scientific "Breeding"

1. Kim Painter, "Doctors Have Prenatal Test for 450 Genetic Diseases," *USA Today*, 15–17 August 1997, pp. 1–2.

2. Susan Jacoby, "Should Parents Play God?" *McCall's* (March 1977).

3. "Medical Ethics and the Stewardship of Life: An Interview with Dr. C. Everett Koop," *Christianity Today* (15 December 1978).

4. Richard M. Restak, *Pre-meditated Man* (New York: Viking, 1973).

5. Quoted in Mark Heller, *Eugenics* (Camden, N.J.: Rutgers Press, 1963), p. 19.

Chapter 10: Reshaping Heredity

1. *Time* (19 April 1971), p. 43.

2. *Newsweek* (20 August 1979), p. 41.

3. *Science Digest* (March 1967), p. 52.

4. Figures from the *1996 Universal Almanac*, ed. John W. Wright (Kansas City, Mo.: Andrews & McMeel, 1996), p. 221.

5. Karen S. Petersen, "High Anxiety Rooted in Brain Chemistry, Genes," *USA Today*, 18 August 1997, p. 4D.

Chapter 11: Where Do We Go from Here?

1. *Time* (19 April 1977), p. 32.

2. Ibid.

3. George Wald, "The Case against Genetic Engineering," *The Sciences* (September/October 1976).

4. Ibid.

5. Ramsey, *Fabricated Man*, p. 96.

6. *Time* (19 April 1977), p. 51.

7. Brody, *Ethical Decisions in Medicine*.

8. Mark Fackler, "Abortion: Two Views," *His* (February 1979), p. 18.

9. For information on the Creation Research Society, contact Dr. Glen W. Wolfrom, P.O. Box 8263, St. Joseph, MO 64508. Or e-mail crsnetwork@ aol.com. For an excellent scholarly book that refutes evolution, see Phillip Johnson's *Defeating Darwinism by Opening Minds* (Downers Grove, Ill.: InterVarsity, 1997). If not in your church or local public library, check your Christian bookstore.

10. *Communication: Ethical and Moral Issues,* ed. Lee Thayer (London: Gordon and Breach Science Publishers, 1973).

11. Jacques Monod, *Chance and Necessity* (New York: Alfred A. Knopf, 1971), 180.

12. These and other themes are postulated by Winston Weathers, professor of English and director of the Rhetoric Program at the University of Tulsa in his chapter "Literary Communication" in Thayer, ed., *Communication: Ethics and Moral Issues.*

Lane P. Lester, Ph.D., has taught at the University of Tennessee at Chattanooga; at Purdue University, where he earned the Ph.D. in genetics in 1971; and at Liberty University. He is presently professor of biology at Emmanuel College in Franklin Springs, Georgia. Dr. Lester was elected to Sigma Xi, the honorary science research fraternity, and is listed in *Who's Who in the South and Southwest*. He is a member of the board of directors of the Creation Research Society and serves as managing editor of the *Creation Research Society Quarterly*.

James C. Hefley, Ph.D., has presented lectures in mass communications for the Staley Foundation at more than twenty colleges and universities and is presently an adjunct professor at Hannibal-LaGrange College in Hannibal, Missouri. He holds the Ph.D. in mass communications from the University of Tennessee, the M.Div. from New Orleans Baptist Seminary, and the honorary Doctor of Letters from Ouachita University. Dr. Hefley has written more than seventy books, five of which are related to science and religion.